THE

DNA

OF

HEALING

THE DNA OF HEALING

A Five-Step Process for Total Wellness and Abundance

MARGARET RUBY

Cover design by Jane Hagaman
Cover art © royalty-free/Corbis

Hampton Roads Publishing Company, Inc.
1125 Stoney Ridge Road
Charlottesville, VA 22902

434-296-2772
fax: 434-296-5096
e-mail: hrpc@hrpub.com
www.hrpub.com

If you are unable to order this book from your local
bookseller, you may order directly from the publisher.
Call 1-800-766-8009, toll-free.

Library of Congress Cataloging-in-Publication Data

Ruby, Margaret.
The DNA of healing : a 5-step process for total wellness and abundance /
Margaret Ruby.
p. cm.
Summary: "The DNA of Healing teaches how to neutralize the negative patterns
handed down through our family lineage and reprogram the DNA with positive
patterns that manifest health, wellness, and abundance"--Provided by publisher.
Includes bibliographical references.
ISBN 1-57174-469-X (5-1/2x8-1/2 tp : alk. paper)
1. Medicine, Psychosomatic. 2. Genealogy--Psychological aspects. 3.
Self-care, Health. 4. Healing. 5. DNA. I. Title.
RC49.R78 2006
616.001'9--dc22

2005027991

ISBN 1-57174-469-X
10 9 8 7 6 5 4 3 2 1
Printed on acid-free paper in the United States

Dedication

This book is dedicated to my loving husband, Steve,
for his amazing ability to bring
clarity, love, and expansion to my world.

Contents

List of Tables and Illustrations

Preface

We Can Change the Stories Written in Our Genes

Each cell of your body contains a mastermind within its nucleus: DNA, the blueprint of life. The code within this long threadlike molecule contains the essential information that regulates the cell's activities and that transfers hereditary traits from one generation to the next.

In the spring of 2003, just 50 years after the discovery of the DNA's double helix structure, scientists finished mapping the sequence of human DNA. They are now racing to extract the details from all the genetic information in our chromosomes. Just as significant, but much less known, is groundbreaking research indicating that our DNA is not a fixed code and that our genes can be affected and even altered by our emotions and experiences.

As a pioneer and educator in the field of healing and how our beliefs and emotions affect our health and our DNA, I too have learned that our genes determine much more than whether we inherit our mother's curls or our father's eyes. The way we react to others, the way we communicate, the way we think and express our emotions, even the pattern of our relationships, health, and finances, are all shaped by our genes.

Often these patterns can be traced back to a source we don't typically associate with heredity—the feelings and beliefs of our ancestors—what I call their "stories." These are locked deep in the memory of our DNA, influencing us in ways we aren't even aware of. In other words, along with our hair color and the shape of our nose, an *emotional history* has been handed down to us through our family lineage.

I like to describe our DNA as a giant storybook. Of the 46 chapters (scientifically referred to as chromosomes) in our personal book of life, 22 chapters were handed down through our mother's lineage and 22 chapters were handed down through our father's lineage. Two additional chapters tell us about our connection with God, or the universal life source. Each chapter is made up of tens of thousands of stories—what scientists call our genes.

The genetic memories and belief systems that make up the storybook of our DNA can support us in many ways, but they are also responsible for passing on negative patterns that keep us from living the life we want. Some stories take effect immediately—when, for example, someone is born deaf or with diabetes. Other stories lie dormant and unfold only when triggered by certain events or experiences, bringing sudden and sometimes unwelcome changes into our lives. Time and time again, I have seen how the stories recorded in our DNA determine the course of our relationships, our health, and our wealth. In essence, what happens to you on your life journey is a result of what is written in the life code of your DNA.

What I have discovered—and what this book is about—is that we can change our code. It is not set in stone. *We have the power to change the stories written in our genes and therefore the power to change our lives.*

This book shares easy-to-use, self-healing techniques that work at energetic, emotional, and vibrational levels to help you release the limiting beliefs encoded in your DNA. These processes have helped thousands of people find relief from a wide range of issues, from emotional stress and troubled relationships to allergies and life-threatening illnesses. *The DNA of Healing* reveals the discoveries, research, and real-life experiences that have led to my present work and it describes five basic steps to resetting your genetic code for total wellness and abundance.

We are living in a time of great transformation. Many of us are shifting from believing that we are victims into awareness that our belief systems create our reality and that we are in charge of our own transformations, including our own healing. In this book, you will learn how to heal yourself—how to quickly neutralize negative programs handed down through your family lineage and how to activate your positive patterns so that you can manifest the wellness that is meant for you.

You have a force living within you that can take you beyond your present beliefs. It is the power to heal. I champion that force within you and welcome you to the next step on your healing journey.

NOTE: When I use the word *heal* in this book, I mean it in a spiritual sense, not a medical sense, even if I do not state so explicitly at every usage. The words *heal* and *whole* come from the same Old English root, *hā*. To be healed spiritually with respect to an issue is to become whole with respect to that issue. For example, healing may involve a cure, but it may also involve being at peace with the fact that there is no cure. If you have a medical condition, please seek the advice of a licensed health practitioner. In no case should any of the statements in this book contribute to the delay or omission of appropriate medical treatment.

Acknowledgments

Although a book with one author appears to be an individual project, the reality is that "no man is an island." My relationships and experiences with my family, friends, coworkers, and the many teachers who have crossed my path are the source of knowledge for this book.

I would first like to thank my husband, Steve; my daughters, Allison and Lauren; and my grandchildren, Alex, Ethan, Aydin, Stryder, and Olivia, for encouraging me to do what I came here to do. Next, a huge thanks to Vera Gadman, Annaliese Kohinoor, Nancy Miller, Jessica Thurmond, and the entire team at PossibilitiesDNA for all their work and dedication in helping people reach their highest potential as well as making PossibilitiesDNA one of the finest seminar companies in the world for personal development.

Huge thanks to Patricia Spadaro for her vision, encouragement, skill, and gift for asking the right questions as we worked together to create and edit this work. Another thank you to my agent, Nigel Yorwerth, for his coaching and for getting the book published.

A second acknowledgment to Vera Gadman for the love and caring she put into the graphics as well as all the PossibilitiesDNA materials. If not for her creations, we would not be where we are today.

Finally, I am deeply grateful to all the PossibilitiesDNA attendees and graduates who are making a difference on the planet as they take their light and love to the world.

Part I

Emotions, Beliefs, and Your Genetic Heritage

1

Your DNA Storybook

How can we consciously create the life we want? What role do our emotions and belief systems play in our health? Where do our belief systems originate and how can we work with them to create healing and a more meaningful and satisfying way of life? My lifelong quest to find answers to these questions has taken me to the frontier of the mind-body connection.

Investigation into the link between mind and body has exploded in the 30 years since Dr. Herbert Benson, president of the Mind/Body Institute and associate professor of medicine at Harvard Medical School, demonstrated that relaxation techniques could reduce stress, lower blood pressure and heart rate, and improve health. Today a range of doctors, scientists, and researchers in the fields of mind-body medicine, quantum physics, psychoneuroimmunology, and vibrational healing are providing strong evidence of the link between our emotions and our health and well-being.

In the midst of this revolution, Caroline Myss articulated a key healing principle in her book *Anatomy of the Spirit,* one that underlies energetic healing techniques from ancient to modern times: Our biography becomes our biology. "Our bodies contain our histories—every chapter, line, and verse of every event and relationship in our lives," she explains. "Every thought you have had has traveled through your biological system and activated a physiological response."[1] We know, for instance, the effect of intense fear or rage—our heart rate increases, we clench our teeth, and our blood pressure rises. Myss says that among the experiences that carry emotional energy into our body's system are past and present relationships, profound or traumatic memories and experiences, and belief patterns and attitudes. "The emotions from these experiences become encoded in our biological systems," she says, contributing to the formation of our cell tissues and becoming stored in our cellular memory.[2]

Another step in our evolving view of the mind-body link came from Dr. Candace Pert, an internationally reputed neuroscientist. In her landmark work, *Molecules of Emotion,* Dr. Pert establishes the biomolecular basis for our emotions and helps us understand exactly how emotions affect health. She found that the major systems in the body form a vast network. What carries information between these systems, linking them together, are neuropeptides and their receptors, what she calls the biochemicals of emotion. These "messengers" are in constant communication with the immune system. In effect, she says that emotions are the link between mind and body.

Based on this scientific research, it is time to transcend our concepts of "the power of the mind over the body," states Dr. Pert. "In light of my research, that phrase does not describe accurately what is happening. Mind doesn't dominate body, it *becomes* body—body and mind are one."[3] We must start seeing our emotions, she says, "as cellular signals that are involved in the process of translating information into physical reality, literally transforming mind into matter."[4] In other words, the mind and body communicate through molecules of emotion.

From the work of pioneers like these, we know that our emotions impact our bodies' cells and tissues and that they influence our health. Yet the leading edge of the frontier goes further, and deeper, still. Revolutionary research is showing that our emotions impact us at the most basic level of our DNA, a finding that has far-reaching ramifications.

Our Genes Respond to Emotion

Geneticists have known for some time that environmental "stresses" can affect genes and cause mutations. In the 1940s, American geneticist Barbara McClintock made an astounding discovery that wasn't fully recognized until much later. In 1983, she won a Nobel Prize for her discovery that genes could change their position on a chromosome in response to stress. In her Nobel lecture, she said that "shocks" to genetic material (anything from accidents within the cell to viral infections to altered surroundings) "forced the genome to restructure itself" in order to overcome the threat. (A genome is the total genetic material of an organism.)

"Our emotions and beliefs, and those we have inherited, affect our DNA. . . . Our genes respond to emotions—for better or for worse."

"The sensing devices and the signals that initiate these adjustments are beyond our present ability to fathom," said McClintock. She encouraged scientists to move forward to determine "the extent of knowledge the cell has of itself, and how it utilizes this knowledge in a 'thoughtful' manner when challenged." She called the genome "a highly sensitive organ of the cell" that is capable of "sensing the unusual and unexpected events, and responding to them."

At the close of her Nobel lecture, McClintock noted, prophetically, that scientists in the future would undoubtedly focus on the genome "with greater appreciation of its significance." In making her discoveries, McClintock had initially worked with plants, but scientists later recognized that the mechanism she had identified—genes moving around on chromosomes in response to stresses—could very well contribute to human evolution by creating new mutations. Just as important, this research showed that *our genetic code is not static* but is affected by stresses in its environment.

As it turns out, new research is starting to prove that stresses in our environment do indeed alter our DNA. A landmark study released in the Proceedings of the National Academy of Sciences in December 2004 indicated that major life stresses can actually damage the telomeres (the sections of DNA at the tips of chromosomes) inside the body's immune cells, decreasing the cells' lives. The study compared a group of women caring for children suffering from serious chronic conditions to a group of women with healthy children. An interesting feature of the study is that the results were strongly related to the *perception* of emotional stress. Women in both groups who felt they were undergoing the highest stress levels had telomeres comparable to someone ten years older than they were.

In a *Washington Post* article on this important finding, Dr. Dennis Novack of Drexel University College of Medicine said the new study showed that mind and body are not separate, that "the very molecules in our bodies are responsive to our psychological environment."[5] While more research is still needed, the study does point to a direct relationship not just between chronic stress and our health but between stress (or emotions) and our genes.

The evidence doesn't stop there. Other scientific breakthroughs, from an entirely different angle, also show the link between our emotions and our DNA. Nationally recognized researchers Glen Rein, Ph.D., and Rollin McCraty, Ph.D., working with the HeartMath Research Institute, have shown that focused, loving feelings and specific intentions altered samples of DNA in solution and produced biological effects in and out of the body. In one study, those who were a part of the experiments were able to cause the DNA to wind or unwind, matching their specific intention. The winding of the DNA helix is associated with DNA repair and the unwinding precedes cell division. In one case, the person being studied was able to affect the condition of the DNA when the sample was half a mile away. As a result of studies like this, researchers have hypothesized, though not yet proven experimentally, that it may be possible through conscious heart-focused intention to influence our cell-level processes and even to change the primary structure of DNA—our genetic code.

These exciting studies correspond to work I have been doing for the past 20 years that shows that our emotions and beliefs—and those we have inherited—affect our DNA. Like these researchers, I have found that our DNA is not a fixed code but a flexible code. In fact, I have found that by using specific techniques, we can replace flawed patterns with new, positive patterns. In effect, negative thoughts and emotions are like the environmental "stresses" Barbara McClintock spoke of; they affect what she called our "highly sensitive" genetic material, which is capable of "sensing the unusual and unexpected events, and responding to them." In short, our genes respond to emotions—for better or for worse.

The opposite, I have found, is also true: Our DNA affects our emotions, attitudes, and behaviors. Scientists are verifying that our genes pass on to us much more than physical traits. In 2001, a team of scientists in Barcelona discovered that a genetic mutation of chromosome 15 makes people more susceptible to panic attacks and anxiety disorders. This tells us that rather than being an imaginary illness or a psychological defect, a phobia can result from a mutation in our genes. In addition, Dean Hamer, molecular biologist and head of the gene structure and regulation section at the National Cancer Institute, says that faith is deeply rooted in our DNA—that we inherit a predisposition to be spiritual. In his book *The God Gene,* he claims that a variation of the gene called VMAT2, which he has dubbed "the God gene," plays a small but key role in the spiritual tendencies that are hardwired into our genes. If phobias can stem from our genes, what other attitudes are a result of a genetic predisposition? If spirituality can be inherited, what other feelings and behaviors are passed on through our DNA?

Beliefs and Healing

After the revolutionary breakthroughs in science and genetics of past years, we are now facing a new and exciting frontier—one that goes beyond exploring how emotions and thoughts affect our health. This new frontier weds energetics, emotions, *and* genetics. It brings together science *and* self-healing. It asks us to confront new questions: Beyond our physical traits, what kind of information is encoded in, and passed on through, our genes? How do our thoughts and emotions affect the sensing devices and signals inside our genes? And how can we use this information to heal ourselves?

These questions spurred me to dig deeper into the dynamics of DNA. I started my journey, as many of us do, with a health crisis that catapulted me onto a path of self-healing and examining the role of my belief systems. Looking back over my life, I realize that I knew all along I could heal myself naturally, starting with something that happened to me when I was ten years old. Growing up as the eldest of four children, I often cooked to help my mother. One Saturday morning, I was in the kitchen frying bacon on a pancake griddle with one hand and making toast with my other hand. I picked up the pancake griddle to move the bacon grease to the sink and the handle broke. The hot grease poured down the front of my body from my hip to my foot.

The pain was so intense, I didn't know what to do. So I screamed and started running down the street. My parents ran after me, brought me back home, and rushed me to the hospital. By the time we arrived, the burnt skin all up and down my right leg had peeled off. I had third-degree burns that were excruciatingly sensitive and painful.

At the hospital, my leg was packed with ointment and wrapped. For three months, my mother wouldn't let me walk on my leg at all because she said that was an important part of my leg healing perfectly. She kept my leg wrapped and padded and she carried me everywhere. I *knew* the leg would heal perfectly because nobody told me it wouldn't. Nobody, including my mother, talked to me about scarring, so I was free of that belief—until I returned to school.

When I rejoined my fourth-grade class, I still had a few bandages on my leg. The little boy sitting next to me looked at me and said, "You're going to be so ugly when they take those bandages off." I just stared at him and said, "What are you talking about?" "You're going to have scars all over your leg and we'll call you ugly," he shot back. Puzzled and terribly upset, I told my mom about it as soon as I got home. She simply said, "That's not true at all. Your leg has healed perfectly. Scarring only happens to people who like scars. You don't." Today I don't have one scar on my leg. I don't have one sign of that burn. Even as a child, the universe was showing me that we *can* heal ourselves and that our belief systems play an important part in that healing.

As an adult, I had to be reminded of my healing potential. I was living the American dream and then all of a sudden I found myself in the hospital hemorrhaging so badly that the medical authorities told me I was going to die. We signed all the papers giving permission for surgery and told the doctors to do whatever was necessary to stop the bleeding. They found a cyst on my right ovary and removed the ovary during surgery. They also decided to remove my appendix, even though nothing was wrong with it.

Right after the surgery, the doctor told me I needed to have a hysterectomy. The operation I had just had was so painful that I didn't ever want to go into surgery again. Only an insane person would sign up for such a thing, I thought, unless she absolutely had to do it to stay alive. I realized then that I needed to look at my life differently and I needed to find a way to heal myself. In essence, the universe was saying, "You need to wake up now and take responsibility. You need to take an active part in creating your life." I started searching for alternatives to Western medicine and discovered acupuncture and herbal remedies, which turned my health problems around. That was my first step in realizing that in some cases it is possible to heal ourselves without resorting to the drastic measures Western medicine tells us we need. At critical times, we may indeed need and should avail ourselves of Western medicine because it can save our lives, yet—as I was learning—alternative medicine and our own belief systems can make a big difference.

Awareness Alone Is Not Enough

After the surgery to remove the tumor, I started to care for my body in a more natural way. Yet I still kept getting hit over the head with one wake-up call after another. Partnerships no longer worked. Relationships fell apart. Something I was unaware of was causing a pattern to repeat itself in my life and I needed to find out what it was.

I explored many avenues and, eventually, began to study the role our beliefs play in shaping the events of our lives. This became my passion and my profession. For several years, I traveled around the country teaching people how to alter their belief systems. The organization I was working for taught that when we change our beliefs, our attitudes also change and we are then able to deal with things more effectively. The seminars we offered were wonderful and they changed many people's lives. They changed my life.

As I would work with participants, however, I found a recurring theme. When people saw how their beliefs were at the core of their difficulties, they felt wonderful. They were delighted to be able to see their problems through a new set of eyes and would leave the program feeling empowered. Yet when I would see them again, they would tell me the same stories. They were still thinking about that divorce, they still hated their job, their children still weren't speaking to them, or money was still a problem. "Why are you telling me these same stories?" I would ask, frustrated. "If you understood and altered your belief systems, why didn't things change?"

I was discovering a fascinating thing: *Awareness alone is not enough to change a pattern.* How many of us have vowed that we would not repeat a certain pattern that we didn't like in our parents, only to find ourselves doing exactly what we said we would never do? Why do we repeat a pattern even when we do not want to?

"Like long-lost family secrets, our ancestral stories—hidden deep in the memory of our DNA—influence us in ways we are not even aware of."

I studied many different modalities and I loved all of them, but none of them resulted in the permanent changes I wanted to see. This phenomenon motivated me to search for the hidden factors that would answer my questions: If awareness and a positive attitude are all we need to live healthier and happier lives, why don't more people change more quickly? Why do we repeat patterns that we want to be rid of? Why is it so hard to change those patterns? If awareness isn't enough, what is?

At that time, I was fortunate to have learned from a series of people working with energy and DNA. One was Gregg Braden, whose pioneering work in the field of physics and DNA created a major breakthrough in my understanding of beliefs and emotions and how they can affect the DNA.

Through many experiences, experiments, and studies, I began to piece together another part of the puzzle—one that was hidden deep in our DNA. I found that we inherit the emotions and beliefs of our ancestors. Programmed into our very cells, these deeply embedded patterns influence our health, wealth, and relationships. Thus the reason we don't change is, first, because the pattern itself is programmed into our cells. Second, the programs come not just from this life but from generations past. Like long-lost family secrets, our ancestral stories—hidden deep in the memory of our DNA—influence us in ways we are not even aware of. These inner programs are like tapes that keep on playing, telling us what to do and what not to do. I realized that if we could find a way to access that patterning within the DNA, we could change the programs playing on the tapes. We could regenerate and rejuvenate our lives.

Core Beliefs Are Passed On through Our DNA

We all have subconscious programs in our genes that include the thoughts, feelings, and beliefs we have inherited. Passed on generation after generation, without our conscious awareness, they deeply influence our lives. An experience that an ancestor had generations ago can actually be the cause of a certain pattern you are repeating today. For example, the raging anger of a grandfather can be passed down to you through a gene, giving you a tendency toward rage yourself. The emotions of a grandmother who died seven generations ago by drowning can cause you to be afraid of water, even though you have no trauma from this life that would make you afraid of water. Unless we identify and change the stories we have inherited, which are the real source of the patterns, our problems won't change.

Say, for example, you decide you would like to be wealthy and you put up notes all over your house to remind your conscious and your subconscious mind that that is what you want. This may very well help you stay in a positive mood. If your DNA program is set, however, to the subconscious belief that says, "You don't deserve to be rich," then no matter how much your conscious mind focuses on "I'm rich," you won't attain that goal. You will keep on repeating your old pattern until you find and remove the subconscious pattern that is sabotaging your new goals. The same is true with positive affirmations. Ultimately, they will only work if your conscious desires are in 100 percent alignment with your unconscious beliefs.

One woman I taught came from an abusive background. Both her parents were alcoholics who abused her emotionally and physically. She ran away from home at the age of 16 and vowed never to repeat that pattern. What happened? She married five different times, all five times to abusive men. She was powerful and beautiful, but when she would get into a relationship, no matter how much she wanted to change, she would start giving away her power to the men in her life and the abusive pattern would repeat itself over and over again. She saw and understood the cycle of abuse in her family and she had decided long ago that she did not want to be around abusive people. Yet her awareness of the issues was not enough to change her life.

When she used the DNA techniques you will learn about in this book, she discovered that the story of abuse came not just from her parents but went back ten generations. One of her ancestors had been falsely accused, thrown into prison, and subjected to terrible abuse. She inherited the belief system that "abuse is part of what you have to take in life and there is no way out." As soon as we found where the abusive pattern originally came from and then removed it energetically from her DNA, she was able to break the pattern of abuse for the first time in her life. She is now in a wonderful relationship.

The Surprising Science of DNA

Science has made tremendous progress in the last several years in understanding DNA, and it is important to be aware of the scientific basics in order to understand energetic DNA healing. First of all, the nucleus of every cell in the body has 46 chromosomes linked together to form 23 pairs. The chromosome has two long, tightly coiled chains, resembling a spiral ladder, called DNA (deoxyribonucleic acid). DNA is a long multiunit molecule containing nature's digital code for life on earth. DNA has been referred to as the blueprint of life, as our chemical code, and as a language. The design of the organism and its molecular components emerge from the information (or language) stored in the DNA.

The four main coding elements, or base units, on the DNA ladder are T, C, A, and G, or thymine, cytosine, adenine, and guanine. When these four coding elements are combined with phosphates and deoxyribose sugar into groups of three, they are called codons. There are 64 possible codon combinations; 61 are used to encode the 20 amino acids, and three "stop" codons are used to indicate the end of a protein sequence.

This coding is similar to Morse code. Morse code uses combinations of two elements, the dot and the dash, to specify letters of the alphabet and punctuation marks. In the genetic code, combinations of the four subunits A, T, C, and G make up the protein alphabet. In Morse code, a brief period of silence specifies the end of the message, just as the three stop codons mark the beginning and end of a protein sequence. When a code becomes altered, it can seriously reduce the function of the protein.

"The fact that we are only using 20 codon combinations but have access to 44 more opens the door to many more possibilities for life that we have not yet seen."

In 2003, scientists determined the complete sequence of DNA in all the genetic material in the human body, one of the greatest scientific adventures of all time. The mapping of the genes was a quest to find the instruction book for how the human body works. Scientists theorized that every function of the human body is controlled by a gene. If they could list all the genes and what function each gene controlled, they believed they would have a map to help them figure out how to change the human body—to prevent or heal cancer or diabetes, for instance. In order for this theory to be true, there would have to be a gene for each of the hundreds of thousands of functions the body performs. When scientists found only around 33,000 genes instead of hundreds of thousands of genes, they were dumbfounded. They had to revise their theory that genes alone were responsible for everything that takes place in the body.

In mapping the genes, scientists found other puzzling facts. Have you ever thought of yourself as related to a banana or even a worm?

Scientists discovered that 50 percent of our genes are like those of a banana and that 40 percent of the worm's genes and half of the fruit fly's genes are similar to human genes. They also discovered that the genes in a newborn baby are 99.9 percent identical to the genes in every other baby being born.

Bryan Sykes, a leading world authority on DNA and human evolution and author of *The Seven Daughters of Eve,* has affirmed the incredible interrelatedness of humans. He created a genetic map of Western Europe and drew the astounding conclusion that almost everyone of native European descent, no matter where they now live in the world, can trace their ancestry back to one of seven women, whom he calls the seven daughters of Eve.

In my work, I too have found similarities in our genetic information. I have found that within our genes we share similar emotions, beliefs, memories, and stories. Universally, we find themes of abandonment, betrayal, loss of trust, separation, and feelings of rejection. We find common memories of droughts, plagues, and famines, of inquisitions, holocausts, and a slave trade. Because of this similarity in the experience of the human species, we can apply processes that involve archetypal patterning to discover what story was handed down through our family lineage.

In their gene research, scientists have learned something about our codons that surprised them. Codons are like computer chips that hold programs of possibilities. While there are 64 codon combinations available to us, scientists found that we are only accessing 20 of them. What happened to the other 44 possible programs? Did they just disappear? Are they sleeping within us, waiting to be awakened? The fact that we are only using 20 codon combinations but have access to 44 more opens the door to many more possibilities for life that we have not yet seen.

The Chapters of Your DNA Storybook

As my work has progressed over the years and as scientists have sought to find out why there are only 33,000 genes in the body, I have asked myself: Could it be that something is blocking our access to all the potential programs in our DNA? How does DNA really work? If genes alone do not control all the functions in the body, what does? Right now, scientists are focused on answering that last question by finding out what turns genes on and off. In the work that I do, I have found that it is our thoughts, emotions, and belief systems that turn our genetic programs on and off, as I discuss in chapter 2. To understand this approach, it helps to think of DNA as a storybook—as our book of life.

First, we must go beyond the textbook image of DNA as simply a double helix or twisted ladder that holds our genetic codes. Instead, picture the DNA itself as a book. Authors like Gregg Braden and Matt Ridley have compared our DNA to a book with chapters. I like to think of DNA as a giant book on tape that lives in the nucleus of our cells. The memory bank of this book holds a huge storehouse of information we never knew existed.

"I like to think of DNA as a giant book on tape that lives in the nucleus of our cells."

Your DNA storybook is made up of 46 chapters, called chromosomes. Each chapter contains stories, called genes. We have thousands of genes, and each gene stores over 80,000 stories. Every story is made up of words, called codons. Since there are 64 possible codon combinations, you could say that each story is made up of 64 words and that the words are made up of the four letters A, T, C, and G—the subunits of the genetic code.

One of the most important features of our DNA storybook is that it is not set in stone. The stories can be changed. In fact, you are always adding to the storybook of your life experiences. When something happens to you, those experiences and emotions get added to the mix of your stories. Traumatic or repetitive negative events create major changes in your story. In the same way, you can intentionally change your story for the better by learning to remove flawed patterns and influences from your genes that have eclipsed your original, divine blueprint. Once the disruptive patterns are removed, your body will return to its inherent, healthy patterning.

Twenty-two chapters of your DNA storybook contain stories that have been handed down through your mother's lineage. These stories contain your feminine essence and your masculine essence from your mother's side. Every event, emotion, and action that happened to the ancestors in your mother's lineage, good and bad, are recorded here, including all the losses, betrayals, and abandonment. In addition, within these 22 chapters are codes about how to nurture yourself, how to nurture others, and how to awaken your creative life force.

Another 22 chapters of your DNA storybook come from your father's lineage. These chapters, containing your masculine essence and feminine essence from your father's side, record all the events and emotions having to do with power, control, responsibility, perfection, and taking your work out into the world. Every event and emotion, including every war, famine, plague, drought, inquisition, holocaust, and injustice that happened to the ancestors in your father's lineage, are recorded here.

Together, these 44 chapters tell the stories of the balance between your masculine essence and feminine essence. Energetically, whether we are male or female, we have both a masculine essence and a feminine essence. They are meant to work together within us as a whole and as a support team. In our genetic lineage, however, the masculine and feminine energies are sometimes off balance, creating stories of pain, fear, and loss within us.

The God Code

In addition to the 44 chapters that have been handed down from your father's and mother's lineages, two additional chapters in your DNA storybook concern your connection with and your separation from God, or the universal life force. The code recorded in these two chapters—which can be called the God code—connects you to the center of your being, where you are one with God and one with the spirit that dwells in all beings.

Our original blueprint contains a deep connection with our Source. Our God code is not always turned on, however. The experiences and beliefs of separateness and duality that have been handed down through our lineage can cause a defect that turns off this code.

When we are trapped in old, inherited stories of scarcity, competition, and separateness, it is hard to access the information stored in our God code. With our God code turned off, we feel disconnected. We see heaven as spirit and the earth as material. We forget that spirit is all around us and that one spirit unites us all.

One belief that has been handed down to many of us is that we need to look outside ourselves for answers. We are taught to see God as detached from us. To think that we are separate in any way from God or from each other is only an illusion of the belief systems we have inherited from our society and our culture.

The cultural belief system that surrounds us can prevent us from connecting with the highest part of ourselves locked in our DNA. Some cultural beliefs are based not on our God code but on materialism, duality, greed, and scarcity—what the "system" says is right or wrong, good or bad. Yet we are so much more than this. We are not the person that our parents or our siblings told us we were—the person that did not listen, the person that was not respectful, the person that talked too much, the person that was not smart enough or good enough, or the person that could never live up to someone else's expectations. We are much more than the results on our report cards or in our bank accounts. We are bright, shining stars who are deeply connected to universal knowledge. When we go deep into our DNA and energetically erase the stories of separation, we can once again unite with our code of perfection.

"We are bright, shining stars who are deeply connected to universal knowledge."

Interestingly, recent research indicates that science may be finding the evidence for the God code. As I mentioned earlier, molecular biologist Dean Hamer believes there may be a genetic link to spirituality. He found that a variation of the gene called VMAT2 was shared more often by those who scored high on a self-transcendence test, prompting him to claim he may have found one of the genes responsible in some way for our feeling spiritual. The test measures characteristics such as a feeling of oneness with everything else in the universe, an openness to that which cannot be proved, and a trust in one's feelings.

The Leading Edge of a New Frontier

The research I have reviewed in this chapter shows that our emotions affect our health—our blood pressure, our immune system, and our heart rate, for example. Evidence is growing that our emotions even affect our genes. But how, and how much? How far back and how far forward does the influence of our emotions reach? If our emotional experiences can alter our genes—and if our DNA determines our phobias and even our spiritual inclinations—isn't it entirely possible that the emotions of our ancestors were encoded into their DNA and then passed on to us through our genes? Why couldn't we have inherited an emotional and behavioral legacy that far surpasses our long-held concepts about what we do and do not inherit? And if negative emotional states, like stress, can damage our genes, why couldn't innovative techniques that shift us to positive emotional states *heal* our genes?

Scientists are now aiming to treat inherited disorders or chronic disease by genetically altering genes and then inserting them into the body. But since our limiting belief systems and emotions are the real cause of our disorders, I have found that DNA techniques that work at energetic levels to remove those limiting beliefs are extremely effective in changing our genetic terrain. To put it another way, when conventional Western medicine looks at DNA, it is looking at the amino acids, the proteins, and the chemical makeup of the structure, whereas energetic techniques work with the principle of vibration and energy flow at subtle levels.

Energetic (vibrational) medicine operates on the premise that the body is made up of energy patterns that are invisible to the human eye and that interface with, underlie, and influence our physical cells, organs, and systems. Techniques as varied as traditional Chinese medicine, acupuncture, homeopathy, therapeutic touch, and electromagnetic and sound therapies all work with the body's energy and energy fields. A growing body of evidence shows that when we alter the flow of energy or change energetic patterns, we see corresponding changes at physical, mental, and emotional levels. It is also possible to work directly with the energy field of our DNA to bring about healing. DNA works like our subconscious mind; it holds stories, emotions, and beliefs that we do not recall in our conscious mind. In order to restore a gene to its healthy state, we have to access the underlying emotions and belief systems that originally created the defect in the gene.

In this book, you will explore ideas at the leading edge of this new frontier, ideas that are as applicable to physical and emotional healing as they are to overcoming blocks in our daily lives and creating the kind of world we want to live in. Each chapter unfolds another facet in the mystery of our DNA, including these core concepts:

- Our genes determine much more than our physical traits. They influence how we think, feel, and react, shaping the course of our health, wealth, and relationships.

- Emotions affect and alter our DNA—and, conversely, our DNA affects our emotions, attitudes, and behaviors.

- The emotions that impact our genes come not just from experiences we had in this life. We inherit the emotional patterns and beliefs (or "stories") of our ancestors. Deeply imbedded in our DNA, these ancestral stories influence us in ways we are not even aware of.

- Lastly, we are not prisoners of our genetic heritage. Our genetic codes are flexible, not fixed. Through simple but powerful self-healing techniques, we can reset our genetic codes and with them the stories of our lives.

2

Your Emotional Heritage

The blueprint of each of our lives that is written in code in our DNA includes a vast storehouse of unconscious emotional information and patterning. Once these patterns are in place, there is little, if anything, we have to do to keep them going.

We know how this works in the body, which is constantly creating new cells but not necessarily new patterns. It takes six weeks for your body to make new liver cells, three to four weeks to get new skin cells, four days to get new stomach cells, and only two days to regenerate cells in your eyes. In fact, our body replaces an average of 2.5 million cells every minute through cellular regeneration, yet the patterns in our body often don't change. For instance, even though our liver cells rejuvenate every six weeks, someone diagnosed with liver cancer in July may still have liver cancer in December. That puzzled researchers until they discovered that the DNA holds the codes to cancer. Although new cells are being created all the time, our cells will replicate with a flaw if we have a degenerative pattern in our DNA.

It is the same with the *emotional* patterns encoded in our DNA. Our emotional heritage, inherited from our ancestors, contributes to the patterns, positive and negative, we see repeated over and over again in our lives. In my work over the years, I have seen that as hard as we may try to change an emotional pattern—whether anger, anxiety, or addictive behavior—those patterns perpetuate until we uncover the real source of the defect, the ancestral stories that are hardwired into us.

Along with other factors, the family "stories" embedded in our DNA can determine if we are rich or poor, skinny or fat. They can determine which of us is prone to depression, which of us is most likely to get divorced, and which of us will be attracted to music or to architecture. It doesn't matter whether we are experiencing blocks in finances, career, or relationships; whatever is working or not working in our lives can be traced to the stories programmed into our DNA code, and the patterns won't change until the inner programming changes. To restore wholeness, we first have to find the story that has created the pattern.

This example shows how it works. One Thanksgiving Day, a curious five-year-old, helping her mother prepare the meal in the kitchen, said, "Mom, I love your ham. Tell me how you make it." "Well," her mother explained, "we get a nice ham and we put some mustard and a little bit of brown sugar and some cloves on it. Then we cut the ends off the ham and we put it in the pan." "But, Mom," the little girl interrupted, "why do we cut the ends off of the ham?" "I don't know," said her mom thoughtfully. "It's the way my mother always made ham."

Since Grandma was there that day, they asked her why she cut the ends off the ham. "You know," she said, "I'm not sure either, but your great-grandma is here and we can ask her why it makes the ham so delicious to cut the ends off." So, they all gathered around Great-grandma and asked her about how to make the ham. She began to explain how she always used brown sugar and mustard and cloves and how she used to cut the ends off the ham. "But, Great-grandma," asked the little girl, "why did you cut the ends off the ham?" "Well, honey," said her great-grandma, "I never had a pan big enough to fit the ham into."

A pattern once established repeats itself generation after generation, even when the conditions that caused the action in the first place are no longer there. In this example, the conscious mind could go back through the generations and take a look at how the pattern got started. If a pattern goes back many generations, however, there is no one alive to help us trace the pattern back to its source.

Activating Ancient Memories

We tend to think that we develop certain emotional patterns or ways of dealing with life from being around our family members or those who raise us and imitating how they act. While this is always part of the equation, it is by no means the whole story. It doesn't explain, for instance, the fear of water some people have when nothing happened to them in this life to cause that fear. If you are really afraid of swimming, there is a good chance you have within your genes a memory of an ancestor drowning. If you have what seems to be an irrational fear, you will most likely be able to trace it to a real event that happened somewhere, sometime, in your ancestral lineage.*

When one of my grandsons was three, for example, our family was looking forward to taking him to see the fireworks on the Fourth of July because he loves excitement. We thought he would love it, but with the first big bang he became terrified, so much so that we had to pack up and leave. When we arrived home, he demanded that we shut the curtains and that everyone hide. I felt that his terror must be coming from a memory in his DNA.

Then I recalled that in World War II my father had been assigned to a ski patrol in the Italian Alps. He would never talk about that period in his life and he forbade me to ask him any questions about it. When my father was dying and was in his last stages of crossing over, he began to relive that time in his life. He talked about the horror of the fighting and about walking through the valleys where thousands had died. My mother told me about the many nightmares he had and I could sense his great pain and his fear of the sounds of war. I remembered that my father had been plagued by phobias, and I realized that the horror he had experienced had created a mutation in the gene that deals with phobias and that this mutation had been handed down to my grandson.

---

*The terms "ancestral lineage" and "past lives" are interchangeable in the techniques I use. If you were to use medical intuitive sight to read a story embedded in the DNA, you would not necessarily be able to tell if the story was from a past life or if the story was handed down through the ancestral lineage. It is also possible that both cases exist simultaneously; that is, you can be born with certain past-life memories stored in your DNA and you can inherit the same or similar stories through your ancestral lineage.

Geneticists talk about genes being "turned on" or "turned off." Some cells express (turn on) certain genes and repress (turn off) others. This is a natural part of "genetic regulation," which makes the cells in our kidneys, for instance, different from the cells in our brain. Scientists are now trying to discover what turns a gene on or off, believing it may open the door to new advances in the field of genetics. Over the years of my work with DNA and healing, I have found that the key factor that determines which genes are turned on or off is the emotions we experience.

Genes are "on" or "off" in the sense that the genetic programs we inherit don't always express themselves immediately when we are born. Sometimes specific patterns don't show up in our lives unless something happens to trigger them—to turn them on. The potential that we would develop a certain illness may have been in our genes all along, but we don't experience it until a specific event or emotion triggers an ancient memory, awakening the gene and causing the illness to come to the fore. Like so many functions in our body, all this goes on without us even knowing it.

In his book *Walking between the Worlds: The Science of Compassion,* Gregg Braden talks about the effect of emotions on our DNA. He cites work by genetic researcher Dan Winter that suggests the possibility that emotion (which Winter says is a long wave) programs the short wave of DNA and that the interaction of emotion's wave form on the double helix in some way determines the site of active or inactive genetic codes. "The implications of this study alone," says Braden, "are vast and profound as we view a possible link between DNA and emotion."[1]

Trauma Triggers Illness

What happened to Sam* is a classic example of how our present-day traumas can trigger illness. Sam was a teenager who suddenly developed an allergy to wheat when he was 16 years old. Prior to that, he had never had any sensitivity to wheat. So serious was this new allergy that eating wheat caused Sam's body to react as if he were dying. As I worked with

---

*The names in the stories throughout this book have been changed.

Sam, introducing him to DNA self-healing techniques, we discovered that the genetic story underlying his problem went back four generations to his great-grandfather. At the age of 16, Sam's great-grandfather had come to the United States from Russia after his family had been massacred in a wheat field. From this horrific experience, his great-grandfather took on the belief (or story) that "wheat is related to death." Because the episode was so traumatic, this new story became part of his great-grandfather's genetic "memory," literally programmed into his genes. The same story was handed down to Sam, embedded in his genes.

"Emotions are, in effect, the vibratory templates that turn our genes on or off. They activate the stories that have been sleeping in our genes."

On Sam's sixteenth birthday something happened that awakened this story. His father died. This trauma triggered the story in Sam's DNA that wheat is associated with death. As a result, not only did he develop an allergy to wheat, but his body also reacted to it "as if he were dying." By applying the vibrational healing techniques I teach, we energetically neutralized and then removed the belief encoded in his genes that "wheat causes death." With that, Sam's allergy to wheat disappeared.

This is a dramatic but by no means isolated example of how our emotions and the belief systems buried deep in our DNA memory can affect our health. Emotions are, in effect, the vibratory templates that turn our genes on or off. They activate the stories that have been sleeping in our genes. This explains in part why children born into the same family do not necessarily develop the same diseases or conditions or grapple with the same emotional issues. Although children from the same parents have the same stories recorded in their DNA, the stories lie dormant until an event from this life triggers the story into action. Whether or not each child manifests their various ancestral options is determined by the emotions that each one experiences. That is what turns the genetic program on or off. For example, my brother has arthritis, but I do not. He had a key experience that I did not have: He attended boarding school away from home, which was a lonely time for him. He felt separated from the family, and this feeling of separation triggered his arthritis.

The good news is that we no longer have to be prisoners of our genetic heritage. The main principle behind DNA healing is that our DNA is not a fixed code but a variable code. In chapter 1, I reviewed scientific research that indicates that our genetic code is not static but that it changes—and that our emotions contribute to those changes. I said that cutting-edge research into the interaction of emotions, genes, the immune system, and behavior has led key researchers such as Dr. Candace Pert to observe that "mind doesn't dominate body, it *becomes* body—body and mind are one." Dr. Ernest Rossi, a psychotherapist and pioneer in the psychobiology of mind-body healing, actually speaks in terms of the "mind-gene connection" and says our daily thoughts, emotions, and experiences can modulate gene expression.

I have been working with energetic self-healing techniques that put these principles into practice. These techniques show that we do have the power to change the stories written in our genes—with profound results. Anyone can learn to work energetically with his or her DNA and return it to its original blueprint of vibrant health and abundance. Once you learn how to discover and decode the core belief system and the emotional source that activates or deactivates your genes, you can neutralize and then remove the old, limiting patterns. You may still have the ancient memory or story within you, but you will no longer have the emotional attachment to it that can activate the story. You can be free of its influence.

Emotions and Our Genes

On one of my trips to Scotland, I experienced firsthand how emotions can trigger specific genes. To understand the background of what happened, you have to know that when my mother was pregnant with me, somewhere around the fifth month, she had a miscarriage. A month later, the doctors were surprised to discover that she was still pregnant—she had been carrying twins. The doctors said

How Emotions and Beliefs Buried in Our DNA Affect Our Health

The genetic programs we inherit don't always express themselves immediately. Ancient memories that have been sleeping in our genes are triggered by the emotions and traumas we experience, causing illness (or some other kind of disruptive pattern) to come to the fore. Once we learn how to decode the core beliefs and emotions that activate or deactivate our genes, we can neutralize and then remove the old, limiting patterns.

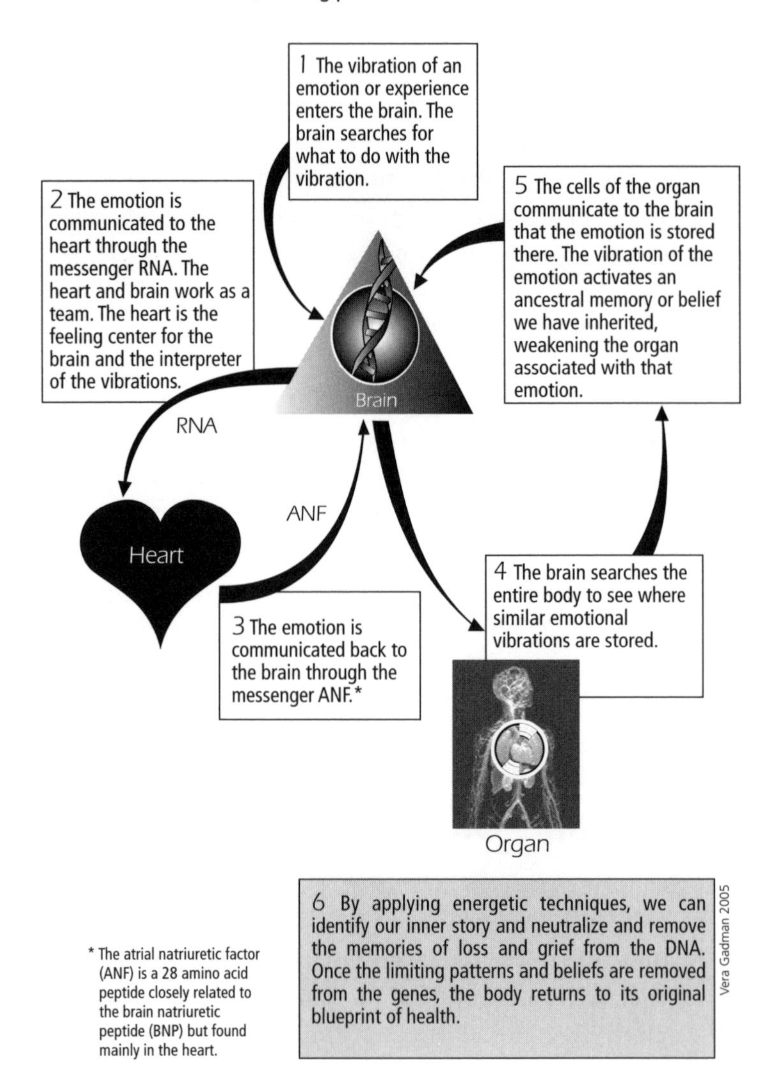

* The atrial natriuretic factor (ANF) is a 28 amino acid peptide closely related to the brain natriuretic peptide (BNP) but found mainly in the heart.

How Emotions Triggered a Severe Allergy

This diagram gives an example of how our emotions and beliefs can trigger illness, using the story of the teenager who, without any prior symptoms, suddenly became severely allergic to wheat (see pages 23–24).

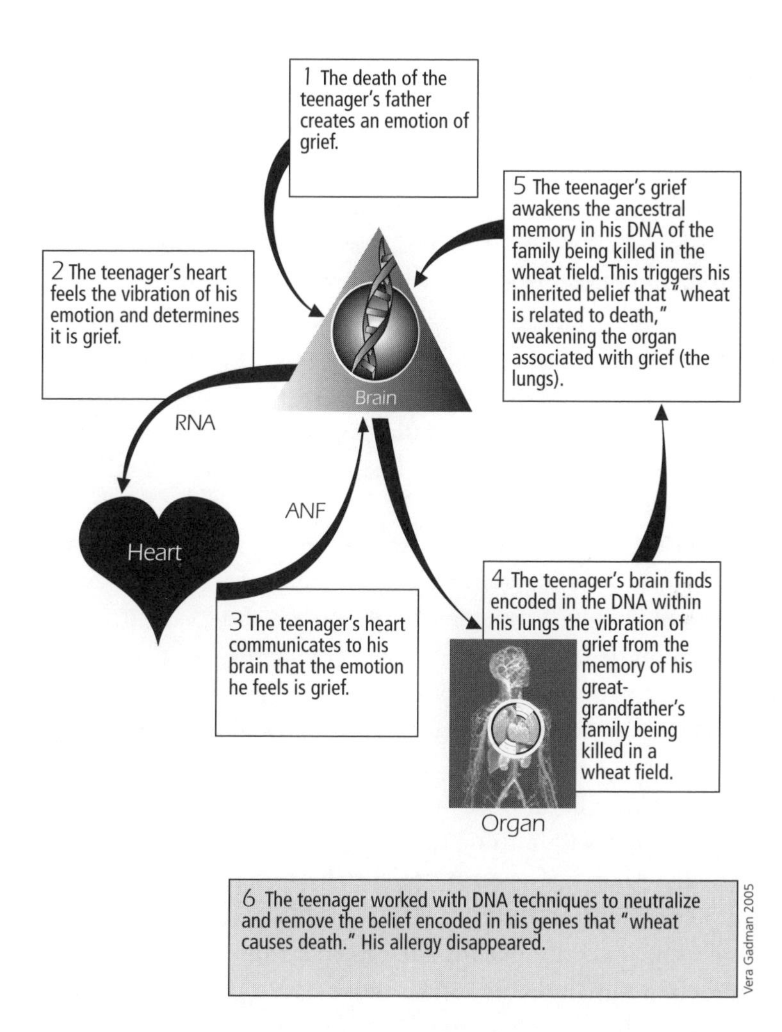

that the medicine they had given her after the miscarriage could be very toxic to a baby in the womb and that they wanted to perform an abortion. After much debate, my mother decided to continue with the pregnancy.

Fortunately, I was born healthy, although there were some complications. After I was born, I developed weakened lungs, a condition that plagued me for many years. Every fall, I would catch a cold and the cold would turn into bronchitis. As soon as I got better, I would catch another cold and it would turn into bronchitis. This cycle would continue until springtime. Until then, I would live on antibiotics and inhalers. I never imagined that all of this would change one day because of a trip to Scotland.

I was visiting Scotland in the early fall and I wanted to see the ancient standing stones in the Kilmartin Valley. It was a cold blustery day, the kind of day where the wind creeps through your bones. In the landscape surrounding Kilmartin, people have lived, loved, danced, mourned, hunted, and prayed for more than ten thousand years. As I stood on the edge of the valley, I could see the burial cairns stretching before me, each an example of the layered history hidden beneath the stones. I could feel the struggles that had taken place in the landscape from the earliest of times.

Standing in the shadow of Carnassarie Castle, which guards the valley, the memories of a time when the land was in terror put a chill through my bones. In reading the history of the castle, I had learned that Duncan of Auchenbreck lost his holding on the castle because of the part he had played in the uprising against the king in 1685. The people who had lived in the valley lost their livelihoods and their cattle, and their sheep and horses could be seized on the slightest pretext. All the memories of betrayal, fighting among the clans, and loss of relationships were held within those stones.

As the wind swept around me, I felt as though I had been in that spot before. Ancient memories of loss and struggle flashed before me. I could not shake the feeling of sorrow. My father's side of the family had lived in Scotland and this was more than a tale from a history book for me—I was remembering an ancient history written in the DNA of my ancestry. The memories of the intricate story of the relationship between the land and its people had been sleeping in my DNA. When I stood in that valley, the memories awoke within me.

By the next morning, I had come down with a cold. By the time I returned to the United States, the cold had turned into pneumonia, but for the first time in the history of my chronic illness I understood what had triggered it—the memories of what had occurred in Scotland. At this time, I was also in the process of losing a relationship with a friend. The emotion of grief that had surfaced in Scotland, compounded by the pain of losing my friend, had weakened my lungs.

Burdened by these emotions, I went deep within. As I looked back over my life, I remembered the story of my birth and the loss of my twin, which was the first pain of loss I experienced in this life. I remembered that both my grandmother and my grandfather had died of pneumonia. As I delved further, using a technique I had developed to reveal the archetypal patterns related to our issues, I saw the same patterns showing up. (These techniques are explained in chapters 4 and 5.)

In the Chinese five-element theory, weak lungs are associated with the metal element and with the emotion of grief—and I was experiencing both weak lungs and grief. As I explain in chapter 4, I associate the metal element with the archetype of the Alchemist and its ability (or lack of ability) to create change and transformation. Putting all this together confirmed for me that the grief from the ancient memories and my grief over my inability to change the course of the relationship I was about to lose had triggered the weakness in my lungs. I knew that in order to heal, the ancient memory from Scotland as well as my present-day grief had to be neutralized. With these key parts of the story in hand, I was able to apply the techniques I had created to energetically neutralize and remove the memories of loss and grief from my DNA. Ever since, my lungs have been strong.

Piecing Together the Threads of Our Story

Richard, a client of mine, recovered from a serious illness by understanding the emotional legacy he had inherited. He had been in and out of a wheelchair for two years by the time he came to me for help.

Plagued by a host of illnesses, including lupus, he had lost most of his muscle mass. The medical community felt they could do no more to help him.

The first thing we worked on together was to find the emotional cause of the illness. Before he had been diagnosed with lupus, Richard had worked as a consultant in a large firm. He had had to quit when he got sick. I asked him about his work and he told me that just before his illness he had been promoted to a demanding position and that it was hard for him to keep up with the demands of his new role. I then asked him if there had been anything unusual about his early childhood. He told me that shortly after his birth his father had died and that his mother had a difficult time dealing with the demanding life of a single mom. He had now used the word *demanding* two times in his story.

"Piecing together the threads of his story and then energetically removing the core emotion from his DNA had a stunning effect."

I took him through another step in the healing process to discover the archetype and emotion behind his disturbance. The process led us to the wood element, which is related to the archetypal patterning of the Pioneer (associated with freedom) and the emotion of resentment. So part of his story went like this: With the demanding work that had been placed on him, Richard felt a loss of the pioneering spirit of freedom. His feeling of resentment at losing his freedom and having these demands placed on him could be traced back to his mother's resentment of having to cope with the demands of being a single mother.

This theme went back even further. Delving deeper into the pattern using the numerical mapping technique (described in chapter 4), we traced the emotion of resentment back three generations on Richard's father's side, where demands were made on his great-grandfather. Richard verified that his great-grandfather had emigrated to the United States as a young boy and that his great-grandfather's entire family had been counting on him to earn enough money to bring them all to America. Thus Richard's great-grandfather had also carried the burden of a demanding situation and felt separated from his family, emotions that were echoed in the loss both Richard and his mother felt at the death of his father.

Energetically, the story written into Richard's DNA, stemming from the time of his great-grandfather, was the core cause behind his lupus. Richard had inherited the belief that demanding situations create a loss of freedom. Almost from birth, he had experienced this story as he grew up seeing his mother struggle with the demands of being a single mother and sole provider for their family. When he was promoted, the same story and all the emotions related to it were triggered again, this time activating a serious physical condition. By developing lupus, he was simply playing out the belief that "demanding situations will make me lose my freedom." In his case, the loss of freedom manifested as lupus, where he no longer had the physical freedom to do what he loved. Piecing together the threads of his story and then energetically removing the core emotion from his DNA had a stunning effect. Within 11 weeks, Richard's body had regenerated his muscles and bones back to health.

How Far Back Do Your Money Patterns Go?

So far in this chapter I've talked about how our emotional heritage can affect our health. Let's take a look at how it can affect another area of our lives: our finances. Say you had an ancestor who found it difficult to make money or to hold on to money and as a result he or she developed a belief that money was hard to get or that one has to work hard for money. That belief about money was literally absorbed by and imprinted into your ancestor's cells. It became, so to speak, a part of that person's cellular memory and was stored within the gene, awaiting future referencing.

How can your ancestor's belief influence your life today? Let's pretend you just won the lottery, yet you still have a belief system written in your genetic code that says, "I must work hard for my money." If you have that story within you and then you find yourself in a situation where you didn't work hard for your money (because you hit it big in the lottery), you will subconsciously revert to and fulfill the preprogrammed pattern. Unless you change the pattern, you will, in one way or another, lose the money that has come into your life, perhaps by spending it all or even giving it away, because you are driven by a subconscious feeling of guilt. You didn't "work hard" for the money, so you don't believe you deserve it. In fact, statistics show that almost 90 percent of the people who win the lottery use up their money within ten years and some lose it within weeks. Why? Because the story written in their DNA wins out.

"Once activated, the belief of your ancestors will play out in your life through the vibration of cellular memory handed down generation after generation. Only you can break the chain."

Of course, many of our patterns with money are a reflection of how we saw our family handle money. If we were raised in an environment where there was plenty of money and money was shared freely, we will adopt the same pattern. If we were raised in a family where money was scarce and family members were afraid of not having enough, we too will be afraid of losing our money or not having enough, no matter how much money we actually have. Yet the behaviors that we observe and learn from our family often have their roots in the events and emotions experienced by our ancestors.

My father's family came to the United States from Scotland and they brought the frugal gene with them. Although there was plenty of money in my father's family as we were growing up, because of the frugal gene they had inherited everyone was extremely careful and didn't spend money freely. It was considered an outrageous waste if you didn't use a tea bag more than once. That frugal belief system goes all the way back to the time of the clans and the memory that they needed to "save for a rainy day" because they never knew what was going to happen or where they were going to be living or what the next season would bring. They lived in constant fear of not having enough food or abundance to survive.

It is, of course, fine to plan for the future, but we limit ourselves if we are always driven by fear. I have worked with clients who were well off financially, but they lived in constant fear of losing their money because a pattern of fear was embedded in their genetic code. Once activated, the belief of your ancestors will play out in your life through the vibration of cellular memory handed down generation after generation. Only you can break the chain.

Changing Our Subconscious Signals

The case of Roger, a young man from England who had been trying to sell one of his houses, is a powerful example of how our ancestral patterns around money can impact our lives. He had two mortgages and needed to cut back on his expenses. When we identified his genetic story, we found that three generations ago, when his great-grandmother moved from Scotland to England, she found it difficult to create a new community. Therefore he inherited the belief that new communities are difficult to create. His conscious mind wanted to sell the house, but his subconscious mind was afraid to. Every time potential buyers would come to look at the house, they would, in effect, hear a little voice saying at a subconscious level, *"Please do not buy this house from me, because if you do I will lose my community."*

You can think of the DNA in your cells as sending out the signal of your subconscious intention. Scientists have demonstrated that our DNA does, in fact, emit a signal, which they call photon light. Our DNA is like a lighthouse that is always broadcasting what we feel, and we read each other's signals. By working with the DNA self-healing process to remove his limiting belief, Roger was able to alter the "signal" he was sending out. The very next day, a TV "makeover" program chose to include his house in one of their shows and, of course, the house sold.

A woman who attended one of my seminars needed exactly $3,000 to go on a trip and to attend a course she wanted to take. I helped her identify the inherited belief system that was telling her this wasn't possible and then remove the block from her DNA. When she got home, she found a check in the mail from a marriage that had ended seven years ago and that she wasn't expecting to see any more money from. The check was made out for $3,000 exactly.

In another case, a young woman attending a seminar who wanted to work on her primary relationship had an extremely quick response to the healing process. She had volunteered to go through one of our sample processes and was on stage. For five years, she had been in an unsupportive relationship. We went to the core of where that patterning was coming from and removed that block from her DNA memory. Then, while she was still on the stage, her cell phone rang and she answered it. Her boyfriend was calling to tell her, "You know, the weather's bad and it's getting dark. I've opened the garage door so you don't have to get out to park the car. And I don't want you to stop at the store because I've made dinner." Her mouth nearly dropped to the floor. In their five years together, he had never done anything like that before.

"You can think of the DNA in your cells as sending out the signal of your subconscious intention."

The change we want to see doesn't always happen as quickly as in these examples. Once you have removed a disruptive belief pattern and you hold the intent that a new pattern is going to manifest in your world, your results will come in their own time—sometimes instantly, sometimes in a month or two months or longer. But they will come.

In addition to the ancestral stories that are hardwired into our genes, other powerful factors influence our DNA. Science is showing that the entire universe is in a state of vibration, and every part of the universe, including us, generates its own unique frequency. These vibrations are sent out into the world, influencing and impacting those around us in ways big and small. The words we speak also have their own unique frequency. Words are extremely powerful. They have the ability to create the most beautiful dream or to cause the deepest pain.

The vibration of the words we heard as children became deeply imprinted in our cellular memory. If someone told us we were stupid or not very smart, we believed those words (unless we had other influences that told us differently) and we did a lot of dumb things to prove that we were stupid. If we were told that we were not good at singing or dancing or sports, we took on the vibration of those words and they had power over our life. Conversely, when we were exposed to words that held a vibration of beauty, we took on the vibration of those words, too. If we were told that we were brilliant or exceptionally good at what we were doing or that we were kind, gentle, and full of compassion, we acted in accordance with that belief and experienced a life of grace and ease. For most of us, it's a mixed bag. Sometimes it's obvious what's in the bag and sometimes it's not. The next chapter takes a closer look at some of the other hidden influences that can have a huge impact on our lives and our DNA.

3

The False Patterns of the Matrix

All of us have patterns in our DNA—higher patterns of our original blueprint and false patterns that block us from reaching our full potential. We enter this world with our perfect blueprint intact and, at the same time, sleeping within this blueprint are genetic memories we have inherited that are not compatible with our perfect matrix of wellness and abundance.

The nonsupportive patterns we inherit from our ancestors are not the only blocks to accessing the original blueprint of our perfect self. We also grow up with many culturally ingrained beliefs that inhibit us and are not native to who we really are. I call this system of beliefs "the Matrix," after the movie of the same name. Strong belief systems are a part of all cultures and they color the way we perceive the world. Many of the beliefs we grew up with support and assist us in our lives, and we want to maintain and strengthen those kinds of beliefs. Other beliefs, some of which we are not even conscious of, do not support who we are and can limit and harm us, even contribute to illness. Those are the beliefs we want to learn how to shift.

The movie *The Matrix* is based on the idea that we are programmed into a world that we believe is real but that is just a program, that we are like robots. The reason I use the word *Matrix* is not that I believe we are robots trapped in this world or that there are vicious villains or terrible enemies that need to be destroyed. What really holds us captive is the set of belief systems and ideas we have bought into. The Matrix is the system of beliefs that we have come to accept as truth—a giant web of ideas that we believe is right or wrong, good or bad. The Matrix is the rules we as a society have created that dictate how we should live our lives.

When we first arrive in this world, we are happy little campers. As time goes by, however, we are introduced to the Matrix in order to survive. Our parents are usually the first to introduce us to accepted beliefs. Then our relatives, teachers, friends, church leaders, and television and other media impress their beliefs on us.

A Giant Web of Ideas

Children are such good imitators. They absorb every thought, feeling, and emotion in their environment. Children don't ask if a situation is right or wrong; they just absorb and record what they see onto their "inner" tapes. The first six years of a child's life are the most programmable. By the age of six, the story you have learned from your culture is fully imprinted within you. Research shows that from birth to age two, children are primarily in the delta brain-wave state, an unconscious sleeplike state that is like an adult being in a hypnotic state. Children between two and four years old are primarily in the theta state, a dreamlike, imaginative state we experience as adults when we are half awake and half asleep. At around age six, the alpha brain-wave frequency predominates, a state of calm that is conducive to creativity and assimilating new concepts. Not until age ten to 12 does the beta brain-wave state become predominant. The beta state is the active, focused state of attention associated with normal adult waking activity.

This research tells us that young children are extremely absorbent and impressionable. They easily accept and assimilate what they see, hear, and sense happening around them. As the saying goes, "Give me your children until they are seven and they will be mine." To put it another way, if we could keep our children program-free until the age of six, they would be free of many of the debilitating, limiting, and false patterns the Matrix tries to impose on us.

The beliefs programmed into our inner tapes cover a lot of territory. Here are just some examples of the beliefs we might have adopted from the Matrix:

- Don't go outside without your coat or you will catch your death.
- Don't climb that tree or you might fall and break your arm.
- Be careful with money because it's hard to make.
- Paying the bills is a struggle.
- Good girls don't speak up.
- Good boys are strong and they don't cry.
- Life is hard.
- It is not good to make mistakes.
- Women are the nurturers.
- Women have less power than men.
- I am selfish when I take time for myself.
- A man should not be emotional.
- I'll be okay when I meet the right person . . . or when I lose ten pounds . . . or when I get that promotion . . . or when I get married.

As we grow older, no one needs to remind us to believe in statements like these because our inner tapes are already programmed with this information. These programs, set to play continually in our heads, reinforce what we should and should not say and do and feel.

The Matrix Does Not Support Individualism

The Matrix has no room for individualism. Its motto is "fit in." It tells us that if we are good and follow the rules, we will be rewarded for our efforts and won't be abandoned. If we decide to step outside the Matrix and ignore its rules, we are told that something is wrong with us and we are excluded. Exclusion is the most painful form of all punishments, and we can get trapped into the illusion of the Matrix by giving in to the fear that we will be abandoned if we do not conform.

Many culturally inherited beliefs have to do with fear—fear of rejection, fear of being alone, fear of success, fear of failure, fear of being out of control, fear of being unworthy, fear of being judged, fear of being abandoned, fear of going against the belief systems of our culture, fear of looking foolish or of making a mistake or of not being perfect. Few of us were supported to look for what was right or good or special about us. Instead, most of us were taught to look for what was wrong rather than what was wonderful about ourselves, others, and the world.

I noticed this negative conditioning taking place when I taught first grade quite a few years ago. For example, the children were rewarded for finding the things that were "wrong" in special pictures they were given to color—pictures of a table with only three legs, a bird with the beak missing, or a child hitting an apple with a baseball bat. The children would get points for coloring in anything that was wrong with the picture. Imagination was not encouraged.

"Few of us were supported to look for what was right or good or special about us. Instead, most of us were taught to look for what was wrong rather than what was wonderful about ourselves, others, and the world."

As I child, I was blessed to grow up in a family that encouraged and empowered imagination and individualism. My mother loved my imagination and she was one of my supporters in creating the world of magic. When I was in the fourth grade, for instance, our class was studying the culture of Hawaii. I thought it would be great fun to have a Hawaiian luau. We did not have a lot of extra money at this time, but that never stopped my mother from letting me have fun. She thought that a luau was a wonderful idea and that we could afford such a party. She told me that in Hawaii they eat a lot of fish, and we had a lot of tuna fish. She also said they eat a lot of watermelon, and we just happened to have a watermelon field across the street from our house. And, of course, they drink a lot of "Hawaiian punch," and we could certainly afford that.

So the date for the party was set. We made hula skirts out of crepe paper and strung flower leis. We moved all the furniture out of the living room to make more room for the party and then my mother put my brother and sisters and me to work painting a mural of palm trees all over the living room walls. It was so magnificent when it was finished that we kept the walls just like that for the next two years.

Other people, however, did not like my imagination. Outside our home, I got in plenty of trouble for using it. During art time, our teacher asked us to draw the most magnificent horse we could. As I was drawing my horse, I looked up and saw the beautiful ponytail of the girl sitting in front of me. I could not have asked for a better tail for my horse. So I cut off some of the girl's hair and pasted it right where a tail should go. I showed it to the girl with the ponytail and she cut some hair for her horse. Other children thought it was a good idea, too, and pretty soon they were all cutting hair for their horses. When the teacher came back to the room, she was horrified. I spent the rest of the day in the principal's office. The Matrix did not support my individualism or my imagination.

How Strong Are Our Illusions?

Research has shown just how strong the Matrix can be. A classic study in social psychology by Solomon E. Asch called "Opinions and Social Pressure," published in *Scientific American* in 1955, tested people's ability to trust in their own knowing. (Another way of saying this would be that it was an experiment to test people's vulnerability to being caught in the Matrix.) Those participating in the study were asked to match the length of a line on one large card with one of three other lines drawn on a second large card. Asch said that in usual circumstances individuals will make mistakes in picking the matching lengths less than one percent of the time.

This particular experiment studied what would happen when a participant was placed in a group where all the others had been coached to choose the wrong line. Under group pressure, the subjects who were in the minority accepted the majority's wrong judgments 36.8 percent of the time. Even when the lines differed by as much as seven inches, some of the people still yielded to the majority's erroneous decision. In other words, the Matrix held more power for them than trusting what they saw with their own eyes.

Another example, reported by Professor John Wilson in 1961, involved a tribe in Africa who were shown a film on sanitation. At the end of the movie, none of the 30 or so villagers could describe the story the film was trying to unfold. In fact, when asked what they had seen, they said, "We saw a chicken," which had appeared for about a second in the corner of one of the frames of the film. Apparently, the concept of a motion picture and of pictures appearing on a screen out of thin air was so foreign to the villagers that they literally could not see the movie being projected on the screen. They weren't looking at the whole frame but were inspecting it for details, and the chicken is what stood out to them. They hadn't yet learned to "see" the frame as a whole.

That is what it is like when we are entrenched in the Matrix. We become trained to see (or not see) certain things, and we keep repeating the same patterns without questioning them. We can't see that there is an incredible world out there beyond the belief system of the Matrix. As a result of our false beliefs and our fear of going outside accepted boundaries, we are apt to put our lives on hold. We come to define ourselves by what we do not want and what we do not want to do—"I don't want to make a mistake" or "I don't want to lose my job." We no longer follow our hearts; we follow the way of the Matrix. Yet the Matrix only robs us of our magical powers and keeps us trapped inside it.

I use the word *trapped* because the Matrix is like a web that entangles us and keeps us locked into its belief systems. The web becomes thick and dense with the many beliefs we take on. You can think of the Matrix as a chaotic energy field that extends about 67 feet in all directions—above us, below us, and on all sides of us. It goes with us wherever we go. We're always within this field of beliefs. It reinforces standards of what is right or wrong, standards for how we should act on the job, standards for relationships, and standards for success. We accept many of these beliefs as "Truth" without examining them. Although this giant web of limitations is an illusion, we believe it is real.

A friend of mine from England helped me understand one of the illusions of the Matrix that we have in the United States: the belief that bigger is better. She was teaching a course where I live in Sandpoint, Idaho. One morning, we went out for breakfast and she ordered pancakes. The portions served in England are smaller than the portions served in the United States, and Idaho is famous for its lumberjack-size pancakes—the ones that cover the plate and are piled high. When my friend received her order of pancakes, she said they were not only big enough to feed an army, but also big enough to lie down in and make a bed.

Her comment made me think about how the belief system "bigger is better" has become part of the American Dream and how it has affected our lives. We super-size everything and we no longer question if we need it or what it is doing to us. We have more space, but as a result we have less intimacy and less community. We have wider freeways but more traffic delays—and quicker tempers. We have more conveniences but less time. We fill up our children's days with more activities, and then wonder why our relationships with them are sometimes more empty. We do large things and we create large things, but we don't necessarily do or create better things. "Bigger is better" is an example of a belief system that is separating us from our true values. Is the belief system "bigger is better" serving our lives or is it robbing us of what we truly want?

"We become trained to see (or not see) certain things. . . . We can't see that there is an incredible world out there beyond the belief system of the Matrix."

We are domesticated into the Matrix programming through our culture, but there is also a relationship between the Matrix and the codes in our DNA. Over time, the beliefs of the Matrix that we have accepted become so ingrained in us, so much a part of us, that they are stored in our DNA. These programs are then passed down to future generations through the genes. I have conducted many workshops in Scotland and it is fun for me to see the programs that get handed down through the Scots' genetic heritage. I have found, for example, that the

Scottish heritage includes a patriotism gene, an ambition gene, and a loyalty gene. In the past, members of a clan had an agreement to remain loyal and, today, hundreds of years later, this agreement has become a program in the DNA. Loyalty, however, can cut both ways. It can help us, but it can also hold us back. We might remain loyal when we really need to move on, just because our genes and our belief structure say that is the right thing to do.

How can we free ourselves from the patterns of the Matrix we've taken on? First, we have to awaken to the idea that there is such a thing as the Matrix, that there are belief systems that we have come to believe in so strongly that we will follow them at any cost. Second, if we want to have full access to the perfect blueprint in our DNA, we have to remove the false codes of the Matrix from our DNA. Research is showing that our cells and our DNA have a communication system. The false patterns of the Matrix are, in essence, disruptive energy patterns that interfere with that communication. We cannot fully access the potential of our DNA communication system if our cellular memory is holding programs from the Matrix. When disruptive patterns are removed, the body naturally returns to its original blueprint of wellness and abundance.

The Matrix Disrupts the Balance of Masculine and Feminine

One way the Matrix can deeply color our view of life is in how it paints the roles of men and women. Most of us have a set concept of what "woman" is and what "man" is, and we see these roles as quite distinct because the Matrix tells us they are. The Matrix keeps us from understanding that, whether we are male or female, each of us has fundamentally both a masculine essence and a feminine essence. Unfortunately, we grow up not understanding the true role of the masculine essence and feminine essence that is innate in each of us and how they are meant to work together as a team.

In chapter 1, I explained that our DNA storybook is made up of 46 chapters and that 22 of those chapters contain our masculine essence and our feminine essence from our mother's side, and another 22 contain our masculine essence and our feminine essence from our father's side. These energies are like a dance of creation, each supporting and feeding the other. The feminine energy brings life into being, births new ideas, and nurtures that creation. The masculine energy creates a structure or foundation for these ideas to flourish. It protects and supports, providing a safe place for the movement and mobility of the feminine energy.

The stories we inherit through our DNA that do not serve our highest potential, whether they are about abandonment, betrayal, loss, or scarcity, have their roots in unbalanced masculine and feminine energy, and the Matrix sustains these false belief systems. Rather than seeing the delicate balance of masculine and feminine modeled in those around us, we often see examples of masculine or feminine energy suppressing its complement. Conflict and wars, for example, are situations where masculine energy has become dominant. This happens when the ego of the masculine essence (whether in man or woman) moves into a mode of scarcity and fights to hold on to and expand what it has, forgetting the universal abundance and creativity that are a part of the female essence. As a consequence of the feminine energy being suppressed, we tend to forget who we are, we forget to listen to our hearts, and we follow the rules and regulations of group consciousness even if it is not good for us.

In today's Western cultures where women are called to serve alongside men in careers away from the home, the Matrix is especially noticeable. Just one example is the false idea that someone has to lose in order for someone else to win, a belief system that is sure to create imbalance in the work environment. The healthy feminine side of us knows that the universe is abundant and that it can and will provide for all of us, but when the feminine essence is overridden with Matrix beliefs of competition, struggle, conflict, and the need to win, the atmosphere of the workplace becomes one of fierce rivalry rather than collaboration.

Masculine-dominant energy is protected by those who have attained what they want and will do anything to keep it. Masculine-dominant energy is also sustained by those who have very little and are afraid to rock the boat, for they fear that if they do, things will only get worse. The problem with masculine-dominant energy is not that it is highly structured, but that it holds on to structure rigidly. We become attached to the idea that structure is good and we become attached to the structures we build. Yet to create abundance, energy needs to flow. So while structure (masculine energy) is necessary, at times it needs to be released to allow a flow (feminine energy) of something new into our lives. If we find it hard to change, this could be an indication that we are attached to structure and therefore our masculine energy is dominant and our feminine energy needs strengthening.

On the flip side, feminine energy can also become overly dominant. Dominant feminine energy creates patterns of overnurturing (or smothering) that keep us from taking our talents and creations out into the world. It overrides our masculine essence, which gives us a sense of direction in life. When it is difficult for us to commit to anything or to stand up for what we need in a relationship, we know our masculine energy needs strengthening. If we smother others, we prevent them from growing up as the unique individuals they are meant to be. We contribute to a loss of self-empowerment that prevents their male essence from taking responsibility for meeting life's challenges.

When our masculine and feminine energy is out of balance in one way or another, the stories in our DNA become stories of pain, fear, and loss. When we allow the masculine energy to serve rather than to dominate and when we allow the feminine energy to nurture but not to smother or resist the necessary balance of masculine energy, our lives flow with grace and ease. By taking the programs of the Matrix out of our DNA through intentional inner work and DNA healing techniques, we can restore balance to the masculine and feminine essence.

The Masks of the Matrix

The rigid roles the Matrix tells us we are supposed to take on can wreak havoc in our relationships. Under the rules of the Matrix, our relationships are based on the "masks" we wear. We may wear many masks—the mask of businessman, teacher, lover, caretaker, protector, mother, father, child, and so on. Problems arise when we start to believe that we *are* the mask and therefore defend the mask and its unbalanced male and female energy.

If we are wearing the mask of a parent or teacher and are attached to the idea that children do not talk back to their parents or teachers, we will always defend our rigid (overly dominant male) position and discipline the child without exploring alternatives or examining our own preconceived notions. As a result, we lose the freedom to say, "Gee, I don't know the answer to that question, but I sure can find out," or "Maybe there is a better way to handle this situation; maybe there is something I'm not seeing clearly." When the structure of our beliefs becomes cemented in place, we lose the ability to move freely between our masculine and feminine essence, between structure and openness. We become cut off from the receptive, feminine energy that opens up to new ways of understanding our children and our relationships.

"When the structure of our beliefs becomes cemented in place, we lose the ability to move freely between our masculine and feminine essence, between structure and openness."

At a time in our lives when my husband and I were being called to grow and evolve, the Matrix threw us a real curve ball. In order to meet the challenge, we had to open to an entirely new way of looking at ourselves and our relationship. Steve and I married under the silent agreement of the Matrix that he would take care of me and provide for me and that I would nurture and support him and our family until death do us part. Our agreement worked fine for the first half of our marriage, when I was in the nurturing phase of my life. I loved being a wife and mom and creating a loving environment for our family to grow in. I took care of their needs and supported their dreams. This agreement felt natural to me and I enjoyed it.

After 20 years of marriage, something shifted. One day Steve came home from work and asked, "What's for supper?" and I responded, "I don't know. What are you fixing?" My response was so out of character for the role I had always played that I almost thought someone else had spoken those words. I was as shocked as Steve to hear them. Suddenly, I was no longer interested in supporting everyone else. I had ideas of my own, things I wanted to do, and dreams I wanted to fulfill. What had happened to the nurturing wife? Where did she go?

The Matrix said this behavior was bad and that if I didn't return to being the nurturing mother, I would be abandoned and replaced with a younger, more supportive woman. The Matrix feels threatened by the power of a strong woman. Thank goodness Steve loved me enough and was willing to work with me to help us both understand what was happening. For our relationship to survive, we had to create a new one—a relationship that was not recognized by the Matrix, a relationship where Steve would nurture me and stand by me as I grew and evolved. I was moving from the archetype of the "nurturing mother" into the "going-out-into-the-world mother." Together we survived that evolution as my focus shifted to bringing my gifts out into the world scene in a career of my own. This shift in our relationship was not easy, but we did it. We moved past the limiting beliefs of the Matrix.

Once we had settled into a new relationship where I had begun to take my dreams and ideas out into the world, we were again challenged to examine the belief patterns we had been taught by the Matrix. One morning I walked into the kitchen to find Steve making peanut butter sandwiches. I asked him what he was doing and he said he was helping my brother, a recently divorced and now single dad, care for his five children. Steve announced that he was going to pack the children's lunches, take them to school, pick them up after school was over, and then help them with their homework. Confused and dismayed, I was the one who lost it this time. If my husband was so busy packing lunches, I thought, who was going to fulfill our silent financial agreements that he would be the primary provider for our family?

Steve was now shifting from the archetype of the "providing father" into the archetype of the "nurturing father" and it was my turn to challenge my ideas of how I thought relationships should play out in the Matrix. According to the Matrix, of course, it was not okay for a man to show and express his feelings or to develop his nurturing, emotional, female side. I was being asked to understand that Steve needed to explore this nurturing part of himself. So we made another shift in our relationship, one that, in effect, switched our roles for the time being. Steve wanted to experience the nurturing side of his personality and I wanted to experience what it felt like to take my dreams to the world. We agreed that we would support each other in this endeavor. It was a great healing for both of us. The masculine essence was being balanced in my world and the feminine essence was being balanced in Steve's world.

Energetically, the nurturing mother (or feminine) has to evolve into the wise woman because energy in a healthy state is fluid, expanding, and always changing form. To remain solely in the role of nurturing mother is unhealthy. Just as the nurturing mother must develop out of her traditional mold so that she can experience more of her masculine essence, so the providing male must evolve and bring his caring, feminine essence into balance. As a result of Steve's and my becoming more balanced, we found a new level of love and appreciation for each other. By moving beyond the limiting belief patterns that dictated what roles the male and female should play, we were able to evolve to a place in our relationship where we both felt supported and nurtured. We now work side by side to bring to the world our shared vision that we all have the right to heal ourselves and our loved ones.

Giving Up the Old Roles

I was called to break yet another mold of the Matrix when the role of mother I was playing threatened my precious relationship with my daughters. I had many ideas of what I thought a good mother should be, and I tried hard to fit into those concepts. One belief I held was that if I did not police my children, they would get hurt or get into trouble. I had created rules that I thought would keep a teenager safe from drugs, from joining the wrong crowd, and from getting pregnant. This worked until my oldest daughter turned 13. The more rules I created, the more she rebelled. She was screaming to become an individual and I was desperately trying to get her to conform. I gave her two choices: She could either become me or she could rebel. She chose to rebel.

Just when I thought there was no hope, I met a teacher who spoke about the importance of relationships that create true and lasting meaning in our lives. I was attending one of his seminars and during a break I confided to him that the relationships in my life were not working. He looked at me and said, "Why don't you give it up?" "Give what up?" I asked. "The role you are playing," he replied. "You are playing the role of wife and mother that your mother taught you and her mother taught her and so on. Give it up." The teacher then asked if I had considered another choice—the choice to let my daughter choose for herself what was good for her and what was bad for her, the choice to let her take some responsibility in creating her life.

At that moment, I realized that my daughter could not have a healthy relationship with me while I was playing the role of protective mother. I saw that I needed to change my role from policewoman to the supportive and understanding mother. Steve and I realized the risk involved in making such a decision, but we also knew that our relationship with our daughters was quickly deteriorating. We took the plunge, even though it was scary. By this time, both our daughters were teenagers. We sat down with them both and informed them that we were giving them their rites of passage into adulthood. From this time forward, we told them, they would be the primary decision-makers in their own lives. They would be in charge of determining what they wanted to create in their world. I told them that I would reserve the right to be the protective mother when necessary so that I could offer my advice and that I would tell them when I was playing that role. But it would still be their job to weigh the advice I would offer and make the final choice. We also told our daughters that they could count on our continual support regardless of the types of choices they made.

This new arrangement was extremely awkward. Our daughters did not trust that we would truly allow them to make their own decisions and they tested us. One day our oldest daughter, now a freshman in high school, announced that she was going to drive to Los Angeles with four older boys to attend a fraternity party. At that point, I decided to exercise the right I had given myself at their rites of passage. I had asked for just one thing: that they listen to all my fears and that I be allowed to offer my advice about their choices. I proceeded to tell my freshman daughter how much her fraternity party idea scared me because she would be in an environment where she might not be able to take care of herself. She listened to everything I said and then stated with determination, "I am going anyway." Steve and I seriously wondered if we had made the right decision. Then two days later, my daughter told me she had changed her mind. She no longer wanted to go.

Our younger daughter had entered junior high school and, upon receiving her rites of passage, she too decided to exercise her newfound freedom. She proclaimed, "I am no longer going to school." She had been struggling with below-average grades and felt that school was intolerable. We honored her choice and she stayed home for several weeks. (Steve was sure we were going to have an illiterate child on our hands.) Boredom eventually settled in, however, and she decided she would go back to school after all. Since she was making all of her own decisions now and was in charge of her life, it was her responsibility to find a tutor to help her get caught up with her lessons. She went to work and her teachers were amazed at her academic transformation on her return to school.

> *"The illusion of the Matrix and the role of protective mother that I had become trapped in . . . almost caused me to lose my daughters."*

As Steve and I released the traditional roles of parents, giving up our need to control our daughters and at the same time becoming clear about the roles we were playing, we saw that our fear of negative consequences was unfounded. Our children never chose to do things that were harmful to themselves or others or to put themselves in difficult or painful situations. With their newly established autonomy, they began to make truly wise decisions in a way that Steve and I never expected. As a matter of fact, they were much harder on themselves than we would have been with them.

The most beautiful part of the rites of passage experience was that, with our daughters no longer feeling judged for their actions, trust grew between us. They began eagerly to seek our advice. The illusion of the Matrix and the role of protective mother that I had become trapped in had almost caused me to lose my daughters. The day I decided to step outside the Matrix, we became best friends, and we still are.

Finding Purpose

Recognizing and dealing with our limiting belief systems can create major changes in many areas of our lives. One of my clients, Shelley, found her life purpose by getting to the root of her belief patterns through her DNA work. Shelley felt as if she did not know who she was. She was a second child born into a strongly opinionated family and she never felt strong enough to voice her own opinion. In particular, she idolized her older sister and decided to do anything that would please her. Shelley adopted the Matrix belief that, in order to survive, she must please others before pleasing herself, and this pattern threaded its way through her life.

After high school, Shelley left home to go to college. Even though she was away from home, she found herself repeating the same patterns. She tried to fit in and conform to what was expected of her according to the academics of a college structure and according to what her parents wanted. She attended classes, studied hard for good grades, and tried to please her professors. Her family wanted her to follow a conventional path of medicine in her schooling and she obeyed.

Yet deep inside, something was missing. She didn't know what she really wanted. She felt as if she were wandering aimlessly through a maze. Shelley was terrified to voice her feelings to anyone because underneath the belief that she had to please others was the fear that if she didn't please them, she would be rejected. As a result, she repressed her true feelings.

When Shelley worked through the process of discovering the stories locked in her DNA, she discovered the story of an ancestor who had been in an environment where her thoughts and feelings were not appreciated. When this ancestor actually did stand up for herself and voice what she wanted, her family rejected her. This ancient memory of the pain of rejection had been triggered in Shelley's life because she was born into an opinionated family. Like her ancestor, she felt that if she voiced her own desires and dared to go against the desires of her family, they would reject her. The pattern of having to please others in order to survive was a belief of the Matrix that she had learned growing up, and it had also been handed down to her through her genetic lineage.

Once this story was removed from her DNA, Shelley lost the fear of voicing what she truly desired. She decided that she did not want to follow the conventional path of medicine but wanted to study alternative medicine. She switched to another school to pursue a vocation that supported her own desires. Today she is a successful alternative healer and her family did not reject her. In fact, they are proud of her.

Letting go of our old beliefs is a process and it doesn't happen all at once. I still find myself confronting old patterns whose turn it is to go. For example, I found that a Matrix belief was holding me back from having a successful seminar in a city where I used to live. Years earlier, my husband and I had lived in Santa Cruz, California, where we had opened one of the first video movie rental stores in the United States and went on to build a successful chain of these stores around the Monterey Bay. During the 1989 earthquake, in less than 17 seconds, we lost everything we owned. We tried rebuilding, but the economy was badly hurt and we did not succeed. This turned out to be a blessing, for it led me to the work I am now doing, which I love.

When we later traveled to the Santa Cruz area to conduct a seminar, a funny thing happened. We often give a free lecture and demonstration before our seminars to acquaint people with what we do and they are well attended. This time, however, no one came to our first lecture and only a few people attended our second lecture, including a woman who poked her head in the door for a few moments and then left. After the lecture, I spoke with one of our daughters and told her that perhaps Santa Cruz was not ready for what we had to teach.

My daughter, however, had a different view. She said, "Mom, maybe *you* are not ready for Santa Cruz." She suggested that I use my own technique to find and remove any blocks within me that were preventing me from having a successful seminar. When I did, I was amazed to discover that my husband and I were holding on to a belief system that we would lose everything we owned if we conducted business in that area again. We immediately used our DNA self-healing techniques to release that belief. Within ten minutes, the phone rang. It was the woman who had poked her head in the door at our free lecture. She said she loved what we had to say and was sorry she had to leave. She owned a radio talk show and asked if we would like to be guests on her show the next morning.

It was a delight being on the show and, that evening, we were scheduled to give one final lecture in the area. When we arrived at the location, there was a line of people wrapping around the block waiting to get in. That night we gave a successful seminar to a standing-room-only crowd, thanks to our having let go of an old belief.

How the Matrix Affects Your God Code

The Matrix—the force that tries to convince us what is right and what is wrong, that claims it has the right to define who we are and who we are not—also has a tremendous influence on whether or not we are accessing the patterns of our spirit in our God gene. I explained in chapter 1 that our DNA storybook includes two chapters that represent the code that connects us with God, or the universal life force. Our God code holds programs of clarity and truth and, most important, the program of cocreation.

Cocreation is the knowledge and power to manifest the world as we dream it. Cocreation is the key program we want to awaken and to access within our God code. When that program is not awakened, we look outside ourselves for our source of creation rather than recognizing that we have the power to cocreate with God. We also see the world through the eyes of separation and illusion. In severe cases, people can become so disconnected from each other and from the spiritual world that they perpetuate wars and atrocities.

The Matrix tries to keep us from tapping into the power of cocreation in our God gene. It teaches us that we are victims and that it is okay to blame others when things go wrong. We are taught to believe that we are victims of everything from germs and arthritis to bad backs and bad luck. We are taught that we need to go through someone else to get to God. We operate from a deep-seated belief that we do not have the right or the ability to control or to design our own lives. We are indoctrinated into the idea that we have to look outside ourselves for the answers when the answers are inside us. We have forgotten how to be the master programmer of our own life.

Rather than teaching us to rely on the intuitive force of knowledge within, the Matrix tells us to rely only on facts. When we convince ourselves that the "facts" show that we do not have enough money or enough ability or the right connections to make our dreams come true, we shut down the positive programs in our God gene. Not only that, we lose access to the positive programs in other genes because the God gene is the master programmer of what I call the gene team.

The programs you are or are not accessing in the God gene influence every other gene. For example, if the program that is turned on in the God gene is a program of separation, we will see ourselves as separate from others and from the world around us. This sense of separation then activates a program of scarcity within the gene that controls our beliefs about abundance. Once a scarcity program is activated, we experience scarcity everywhere we turn—there is not enough time, not enough money, not enough food, and not enough love. When we are able to access the God gene's perfect blueprint and its programs of connectedness to all things, all this changes. We are able to see the world through the eyes of truth, and our powers of cocreation blossom.

"We are indoctrinated into the idea that we have to look outside ourselves for the answers when the answers are inside us. We have forgotten how to be the master programmer of our own life."

Any program that says we are separated from each other or that what happens to one of us does not affect the whole of humanity actually creates a flaw within our God gene. Any program that tells us that something is wrong with us or that we do not deserve to have what we love also creates a flaw within our God gene.

Stages of DNA Healing

We have all experienced the crippling effects of the Matrix, whether they show up in our career or our relationships, our finances or our health. We have all taken on patterns from the Matrix that have blocked us from fully accessing our God gene. The wonderful thing about our DNA is that we *can* change it. We can remove the flaws. Like the people described throughout this book, you can learn to identify the belief systems you have adopted that may be causing you to act out the same patterns over and over. You can learn to remove the unsupportive stories that are not serving you and you can learn to add new stories that expand your life. We *do* have control of our world and it starts with asking: Am I going to continue to let the Matrix rule my life or am I going to consciously create the life I want?

"We do have control of our world and it starts with asking: Am I going to continue to let the Matrix rule my life or am I going to consciously create the life I want?"

The foundational process of DNA self-healing involves five basic steps or stages that help you access your inner story and change it.

Step 1. Discover the Theme of Your Story—how to find the overall theme and disruptive pattern that created pain, loss, and hardship and interfered with your having the life you want. The theme of our story comes from the Matrix belief systems we have adopted and from the strategies, roles, and archetypes we have taken on to play out that story. The Matrix beliefs that we take on trigger the original source of the interference patterns in our DNA—the patterns we inherited from our ancestors.

Step 2. Discover Your Ancestral Story—how to find the original source of the disruptive pattern, and the strategy and program your ancestors took on. In this second step, we learn to identify: the stories that have been handed down to us from our family lineage through our genes; the strategy and programs our ancestors took on that helped create the story; and the feelings those ancestral patterns have set up in us.

Step 3. Discover How These Stories Play Out—how the ancestral patterning and themes have repeated themselves in your life. The third step helps us become aware of how the overall theme of our story (a combination of our ancestral patterns and Matrix patterns) has repeated itself in our life and how it affects us today.

Step 4. Create a Change of Heart and Neutralization—how to neutralize the emotions and feelings that created the patterns of disturbance. In Step 4, we go deeply into the heart and bring the energy that is locked in the old patterns back to a neutral state so that we can create a new story through manifestation, which is Step 5.

Step 5. Create Your New Story—how to manifest your new story through the power of intention, imagination, and focus. In this step, we learn how to align our minds, our hearts, and our thoughts with the powerful energy of creation to bring what we want into our lives.

Now you have an idea of the foundational concepts. The next chapters show how numbers and archetypes can help you quickly identify your deep-seated patterns, how the heart plays a key role in neutralization and healing, and how you can use intention and imagination to create the new patterns you want to bring into your life.

Part II

The Language of Healing

4

The Language of Archetypes and Numbers

Many of the patterns ruling our lives are hidden deep within our subconscious, handed down to us from the belief systems of our ancestors. *Subconscious* seems like a big word, but it is not all that mysterious. I describe the subconscious as simply the thoughts and emotions we can't hear but that play over and over within us like tapes, shaping the way we see, interpret, and react to the events in our lives. Our DNA is the system that controls these tapes. We all have subconscious tapes playing all the time, but they don't have to be stuck on autoplay. We can create new tapes, in effect recreating our lives.

In order to change what the tapes are playing, we first need to bring what is in our subconscious into our conscious mind. Why is this important? Many seminars, programs, and books teach us how to get to the root of our problems and tell us that if we can just hold the vision of what we want and get excited about it, it will come our way. Sometimes this works, but often it doesn't because somewhere deep within our DNA is a program we're not even aware of that blocks that vision from coming into manifestation. In fact, it is sending out the opposite message. No self-help technique, even the most incredible, will work permanently when deep-seated patterns are programmed to keep on rolling.

For years I had been studying what kind of thoughts, beliefs, and memories cause us to get sick and hold us back from what we truly want to do and to be. I realized that the story of our life is a combination of several things—the belief systems we have been taught by those who raised us *plus* the conglomerate system of beliefs and rules our society has taught us to accept as truth (the programming of the Matrix) *plus* the stories of our ancestors embedded in our DNA. The strategies and roles we have taken on and the emotional patterns that repeat themselves throughout our lives stem from these core patterns.

I also knew that it is not easy to identify these core patterns on our own. Sometimes it can take years in psychoanalysis to get there. Even then, when we try to uncover those patterns, we are working with our conscious mind. We can talk about the things that happened to us that we can remember, but we don't have access to the stories from our ancestry that we have no conscious awareness of—and getting back to those stories is where real transformation occurs. One missing component of healing is finding the source of the genetic patterns and then removing it permanently. Our behavior doesn't shift—and our life doesn't shift—until we shift our inner program.

How can we identify something that is buried so deeply within us that we don't even know it exists? That was the question. I knew that to identify hidden patterns I needed a map that would lead directly to the limiting emotions, thoughts, and belief systems. The first piece of the puzzle came as I explored the archetypal patterns associated with the ancient Chinese five-element theory and the Indian system of the chakras.

An archetype is an original model that other things are patterned after. Carl Jung enlarged the meaning of archetype to encompass thought and behavior patterns and images that exist across cultures and are universal. Jung believed that these forms and images of a collective nature exist in humanity's collective unconscious and that we inherit them from our ancestors. Archetypes, therefore, are patterns that are both ancient and universal. They have their dawn in human history—and, as we know, history has a tendency to repeat itself.

In my ongoing work with DNA self-healing, I identified five key archetypes based on the ancient wisdom of the five-element theory: *the Philosopher, the Pioneer, the Wizard, the Judge,* and *the Alchemist.* In effect, these archetypes are energetic patterns that influence how we view and react to the world around us. Just as the events and emotions we experience affect our DNA, so events and emotions shape the expression of the archetypes within us. At one time or another, all the archetypes are operative within us, but in certain phases of our lives one or more may be prominent.

I found that if certain archetypes tested positive for someone, it meant that these archetypes revealed parts of the story hidden in that person's DNA. In other words, I found that the archetypes derived from the five elements provide symbolic language to interpret the stories written in our genetic code. They are key parts of the map that lead us to our inner programming.

"The archetypes derived from the five elements provide symbolic language to interpret the stories written in our genetic code."

At first I incorporated kinesiology (or muscle testing) to identify which archetypes were the right ones for each individual. Kinesiology is a science that gives you true and false answers as you resist pressure applied to your muscles (usually the arm). If your muscles are strong when pressure is applied, whatever is being tested is true for you. If your muscles are weak when pressure is applied, what is being tested is false or not good for you. This system had its drawbacks, however. I found that if you were not 100 percent accurate in your testing or did not trust yourself to get accurate answers, the results would not be correct.

I asked to be shown a faster and more efficient way of reading the map of the thoughts, emotions, and belief systems operating within us. A breakthrough came when I was inspired with a simple but profound way of reading and decoding the subconscious programming and stories encoded in our DNA—through numbers. I discovered that no matter how deeply buried the various parts of our inner story are, we can quickly access the story lines by choosing numbers that have been assigned to the archetypes. This system is based on energetic, or vibrational, healing principles.

A Vibrational Map

All our beliefs and emotions have vibrational patterns associated with them. When numbers are assigned to a variety of archetypal human thoughts, emotions, and belief patterns, by the law of vibration each assigned number takes on the vibration of the words or sentence associated with it. The numbers represent the geometric patterning of the beliefs at a vibrational level. In the self-healing processes I have developed, you pick a series of numbers in answer to a series of questions. Because we are all energetic and vibrational beings and because like attracts like, the numbers you choose will unerringly match the vibrational energy patterns your DNA is sending out. Thus the numbers will lead you directly to the core belief system at work within you.

Your body and its hidden belief systems work like a set of matching tuning forks. If you strike one tuning fork so that it vibrates, the other one will begin to vibrate on its own. Likewise, if you have an emotion of grief inside your cells and I ask you to give me a set of numbers to indicate a particular emotion or theme or belief system, you will automatically choose the number that matches the emotion and its corresponding vibration inside your cellular system. You may hear the number in your head, you may "see" it, or, as some people report, the number will just pop out of your mouth before you are even conscious of it. You cannot choose the wrong number because your body is talking. The DNA in your cells is communicating and the number associated with the statement or archetype that matches your own core pattern will come to the conscious level.

In essence, you can piece together your own story with numbers and archetypes as the map. This system of numbers corresponding to archetypes, when used correctly, is a step beyond kinesiology. Kinesiology can be limiting because of the drawbacks I mentioned earlier, and because it is a slower process and practitioners cannot do this kind of work over the phone. The numerical mapping system is fast, effective, efficient, and simple to use. Sometimes it takes just one session to find the core vibration that started the disturbance in the first place and to remove it energetically from the DNA. Sometimes it takes a little more research to find the pattern and remove it. The advantage of this system is that you don't have to have a degree in science to become proficient at it. Healing professionals, such as chiropractors, acupuncturists, and nurses, practice these techniques as an adjunct to their own work, but so do moms who want to heal themselves and their children.

For example, a woman who attended one of my workshops used the process to help her daughter, who was an environmental engineer and was having trouble finding a job because the government wasn't releasing enough funds. When the mother returned home that night after the workshop, she went right to work using the numerical mapping process. Her daughter picked a group of numbers in answer to a series of questions. These numbers, which were correlated to specific archetypes, would reveal the hidden belief system within her daughter that was blocking her from getting the job. Then they used the rest of the process to remove those blocks energetically. The very next day, her daughter found out that the government was releasing the funds and that she would be hired for the job.

In another class, a man who had two high-profit real estate deals in the works learned to use the process. In the past, when it came to closing such deals, something would always go wrong. This man felt that somehow he was sabotaging himself, but he didn't have the faintest idea how. By using the numerical mapping process, he identified what was blocking his success and he released the block. After the workshop, he called us and was excited to report that both of the deals had gone through without a glitch.

Identifying the underlying core belief patterns that are behind the disturbances in your life brings these patterns out of the subconscious and into your conscious awareness. This is the first step to healing and transformation. It allows you to detach from the patterns and look at the events and issues in your life more objectively. With the numerical mapping system, you can identify the key theme that runs through your life (a combination of the strategies, roles, and archetypes that drive you), the Matrix belief systems you have adopted, as well as the stories of your ancestors that were passed down to you through your genes. You can learn exactly where in your ancestral lineage the pattern came in, what the original belief was, and what emotion or event in your life triggered and activated this pattern in you.

Finding the Stories That Sabotage Us

One of the most dramatic healings I ever witnessed using numbers and archetypes to identify, neutralize, and heal issues took place with a friend who now plays a key role in my teaching program. Unexpectedly, she was diagnosed with scleroderma, a rare autoimmune disease that creates an excessive buildup of collagen in the skin, causing swelling, stiffness, and extreme pain. It infiltrated her internal organs, wreaking havoc and threatening her life. "My lungs were working at only 50 percent capacity," she recalls. "I became winded by simply walking from one end of the house to the other. There was fluid on my heart. Both my liver and kidney functions were affected. My esophagus was also involved, making eating a challenge. My hair fell out by the handfuls." In addition, her body was so stiff that she had to shuffle to walk and was unable to care for herself. "I couldn't dress myself, cook for myself, drive, or pick myself up when I fell, and I required 14 to 16 hours of sleep a day."

The specialist who first gave her the diagnosis said there was no known cure for scleroderma. "It takes two to four years for this disease to run the worst of its course. You will probably never have a completely normal, healthy life again," he told her. She says, "I felt as if I had just been handed my death sentence. Strangely enough, the inner core of my being, my spirit, felt perfectly at peace, calm, unruffled."

When she called to tell me about her diagnosis, we went to work right away to find the root of this disease and what had triggered it. As we went through the numerical mapping process and followed the trail of the numbers she picked, we found that as a result of an extremely abusive childhood, she had taken on the belief that she could never have what she truly wanted. This pattern, we learned, had come in seven generations ago, at the time of the potato famine in Ireland. Her ancestors at that time could not have what they truly wanted. Today, however, she did have what she wanted. She was an excellent teacher and loved her work. She had a great marriage. Her husband was building their dream house. Having a happy world, however, was a foreign vibration to her body. It didn't match the ancestral belief she was holding that "I can't have what I want." As a result, her body took on an illness to create a world that was more like what she was used to—a world that would match her belief system.

Once we discovered the emotion and the belief system that had created the energy interference pattern within her DNA and triggered this disease, we neutralized the emotions and took the unsupportive belief system out of her DNA. Instantly, her body began to heal itself. The physicians could not believe it. They had never seen this disease reverse itself. When she later went for a lung test to check her breathing capacity, the nurse thought there was a mix-up with the records because she said there was no way that such a reversal could have taken place. My friend was completely healed. Today she has no signs of scleroderma anywhere on her body or inside her body. Most of all, she's finally living in a world where she can enjoy doing what she loves and have a healthy body, too.

> *"As we . . . followed the trail of the numbers she picked, we found that . . . she had taken on the belief that she could never have what she truly wanted."*

Looking back at this intense initiation, my friend says, "In actuality, my illness contained an awesome gift. The experience propelled me to new levels of insights and growth that I believe would not have occurred without the experience of severe illness. I learned how imperative it is to work with issues as they arise. My recovery required a few months of intense work on myself to accomplish the high level of wellness that I experience today. It required that I challenge myself to enter the deep, dark recesses of my subconscious to pry loose the once-hidden fears and unresolved pain that had evaded my conscious mind for so long.

"Once these patterns were revealed to the light of awareness, the DNA energy healing techniques gave me the means to heal my unresolved fears and beliefs. Each time I used them, I noticed a reduction in the severity of my physical symptoms. Each time I cleared an issue or belief, I felt lighter, freer, and more energetic." In addition to the DNA self-healing techniques, she also held a strong intention to live in a healthy body and continually called on her imagination as a tool, visualizing how it would feel to move in a healthy, vibrant body (a tool covered in chapter 6). "This strong intention to heal sustained me in trying moments and kept me focused on what I desired rather than succumbing to the powerlessness of fear and despair," she says.

The Elements, Their Archetypes, and the Circle of Life

The Chinese five-element tradition is 5,000 years old and in its totality is complex. I incorporate only a small part of it to assist in identifying which element is out of balance, the specific organ or organs that are holding a low vibration, the emotion involved, and the major archetype being played out in an individual's life.

The five-element theory is based on the understanding that everything is connected and everything is a form of vibration. The Chinese divided the energy of these forms of vibration into five categories: metal, water, wood, fire, and earth. They progress in a clockwise cycle known as a "birthing cycle," each one giving birth to another—water birthing wood, wood birthing fire, fire birthing earth, earth birthing metal, and metal birthing water. This circle symbolizes the continuity of life. All things in life go through these stages. The birthing cycle also integrates human emotions with the natural rhythms of the physical body and serves as a map of the stages of growth and decline inherent in all life processes.

As I said, to help us understand the underlying issues we face, I identified archetypes that correspond to the elements. When the five elements and their archetypes are working together in harmony within us, we see their healthy, supportive, balanced expression. The reverse can also be true: We can observe an out-of-balance expression of the elements and their archetypes in our lives. When this is the case, we can learn two things. First, an unbalanced expression of an archetype can show us where we need to make adjustments to bring greater balance into our lives and, second, it can indicate that there are underlying stories in our DNA that need to be neutralized. Through discovering the archetypal patterns that have been governing your life, you will come to understand why you repeat certain patterns and how you can create healthier patterns. You can think of archetypes as guides that show you where you have been and that help you create the potential of where you are going.

The wise ones working with the five-element theory thousands of years ago also discovered that the elements affect specific organs, each of which corresponds to certain emotions. According to the five-element theory, an out-of-balance emotion upsets the health of its corresponding element and organ. Conversely, when an organ is out of balance, unhealthy emotions can manifest. By releasing unhealthy emotional patterns, we can heal the corresponding organ. I have found through my work that the emotional pattern is actually stored in the DNA of the afflicted organ and its related organs and that the emotional pattern can be easily uncovered using the five-element theory as a road map.

The charts on the following pages summarize how the elements, organs, emotions, and archetypes are interrelated and what kind of patterns you are likely to see when an element or archetype is in balance or out of balance. If you are experiencing the physical or emotional symptoms listed in the chart, this may be a sign that this archetype is directing your world.

The five elements and their archetypes, while each performing their own functions, are interrelated and interdependent. The following story shows how the five elements and their archetypes work together in their cycle of action.

Once a great philosopher had a wonderful idea of how to transform the world. He knew he could not create this change alone, so he enlisted his friends the Pioneer, the Wizard, the Judge, and the

The Five Elements

Water Element		
Archetype	Philosopher Understands the bigger picture of the cycles of life. Steps outside the Matrix to create a new vision. Represents renewal and death as the beginning and ending of a cycle.	
Positive, balanced expression of element and archetype	Deep sense of introspection. Ability to remain true to basic nature. Great sense of thoughtfulness and originality. Sage.	
Unbalanced expression	Fear of survival, not knowing what is safe, unable to see the bigger picture.	
Organ and related emotional imbalance	Kidney: fear, terror, dread, bad memory, escapism	Bladder: irritation, timidity, paralyzed will, miffed, inefficiency
Related physical imbalance	Kidney: ▪ dry mouth; ▪ lung congestion; ▪ kidney, low back pain, hip pain; ▪ disc protrusions, ruptures; ▪ knee/ankle pain or soreness, foot/ball of foot pain; ▪ incontinence; ▪ hemorrhoids, hernia, tailbone pain	Bladder: ▪ incontinence, painful urination; ▪ kidney/bladder infections, dropping bladder; ▪ spinal or low back pain or stiffness; ▪ hip and knee pain; ▪ leg ache; ▪ foot pain
Disruptive patterns create loss through	Loss of the ability to see the bigger picture	

Wood Element

Archetype	Pioneer Forges into new territory. Represents birth and growth. Explores the inner self and is not afraid to try new things or to take risks in his or her growth. Includes the inventor and entrepreneur.	
Positive, balanced expression of element and archetype	Driving need for action and fulfillment. Ability to create movement and excitement. Zest for life. Fun to be around. Ability to control emotions while expressing basic feelings. Full of flowing and inspirational energy. Visionary with great creative potential. Needs to create and share beauty. Ability to share enthusiasm, vision, and sense of adventure.	
Unbalanced expression	Force, stubbornness, prolonged anger, overdoing and overperforming.	
Organ and related emotional imbalance	Liver: anger, frustration, rage, irrationality, aggression	Gallbladder: resentment, stubbornness, depression, emotional repression, indecision
Related physical imbalance	Liver: ▪ eye irritation; ▪ lump in the throat; ▪ muscle or joint stiffness, back or rib pain, knee problems; ▪ uterine or prostate complaints; ▪ arthritis; ▪ skin ailments: eczema, psoriasis, dermatitis	Gallbladder: ▪ headaches; ▪ lower leg or ankle pain, muscle pain; ▪ gallbladder pain, gallstones; ▪ tightness in ribs or thorax; ▪ hip or joint stiffness; ▪ digestive problems, belching, gas
Disruptive patterns create loss through	Loss of the ability to move forward	

Fire Element				
Archetype	Wizard Ignites and fuels the spark of life. Creates magic and miracles. Transforms reality with passion.			
Positive, balanced expression of element and archetype	Passion, enthusiasm, optimism, and ability to transform. Enthusiasm to bring transformation through the power of love. Instrument of magic and imagination. Great capacity for joy, affection, and expression.			
Unbalanced expression	Loss of passion, loss of love, suppression of emotion, hopelessness.			
Organ and related emotional imbalance	Small Intestine: vulnerability, sense of abandonment, feeling of being insecure and deserted, absentminded	Thyroid-Adrenal: confusion, paranoia, muddled thinking, emotional instability	Heart: frightfully overjoyed, frightfully sad, shock, guilt, inappropriate laughter, rapid mannerisms and speech	Male/Female (Pericardium-Circulation): depletion, suppression, unresponsiveness, sluggish memory
Related physical imbalance	Small Intestine: ▪ ear problems, tinnitus, sore throat; ▪ digestive difficulties; ▪ abdominal pain; ▪ lateral shoulder pain, length of arm pain and stiffness; ▪ Crohn's disease	Thyroid-Adrenal: ▪ colds, ear infection, tonsillitis, sore throat, swollen glands; ▪ eye problems, cataracts, pink eye; ▪ cold hands/feet, excessive sweating, hot flashes, easily chilled	Heart: ▪ headache, face pain, dry mouth, nausea; ▪ heart palpitations, chest pain; ▪ arm pain/ache, numbness, neck stiffness/pain; ▪ mid-back pain or stiffness;	Male/Female (Pericardium-Circulation): ▪ eye problems; ▪ hot flashes, sweaty palms, poor circulation; ▪ hunger, thirst, sleep disorders; ▪ rapid heartbeat, arteriosclerosis; ▪ depression, mood-

		▪ jaw pain, shoulder/arm/wrist stiffness; ▪ low energy, chronic fatigue; ▪ immune deficiency syndrome; ▪ menstrual cycle irregularities	▪ digestive disorders, constipation, food allergies; ▪ swollen ankles, varicose veins	swings, poor memory, poor concentration; ▪ stroke
Disruptive patterns create loss through	Loss of passion			

Earth Element

Archetype	Judge Mediates and brings all things back to a state of balance. Balances justice with compassion.	
Positive, balanced expression of element and archetype	Capacity for feeling centered, balanced, and at peace with the world. Love of family and home. Ability to care for self while caring for others. Supportive and reliable. Connected and filled with love and compassion for all things. Good mediator and able to infuse balance, gaining people's trust and appreciation.	
Unbalanced expression	Lack of equal energy (feeling that one is physically or emotionally giving more than one is getting), difficulty handling life, loss of individualism, not standing up for self.	
Organ and related emotional imbalance	Spleen/Pancreas: low self-esteem, hopelessness, living through others, overly concerned, lack of control, worry, distrust	Stomach: disgust, despair, overly sympathetic, nervous/stifled, expanded sense of self-importance, obsession, egotism
Related physical imbalance	Spleen/Pancreas: ▪ lymphatic congestion; ▪ loss of appetite, nausea; ▪ abdominal pain, distention; ▪ stomach and/or pelvic problems, female imbalances; ▪ limb fatigue; ▪ leg, knee, or thigh pain	Stomach: ▪ headache, sinus or jaw pain; ▪ stiff neck, swollen throat; ▪ tightness of chest; ▪ stomach, digestive, and gastrointestinal problems (hunger pangs, gastritis); ▪ hiatal hernias, ulcers; ▪ pelvic/thigh pain, knee and shin problems
Disruptive patterns create loss through	Loss of equality	

Metal Element

Archetype	Alchemist Transforms the old, including old beliefs, into new forms and realities. Produces results beyond normal expectations.
Positive, balanced expression of element and archetype	Competent. Has a firm foundation and ability to create essential structures. Great sense of direction for where life is going. Well-organized. Ability to take in the richness of life. Knowing at a deep level that one is already perfect. Inner concentration, mental clarity, and deep understanding for the transitory nature of life. Lives with integrity and principle. Great detachment and discernment.
Unbalanced expression	Being stuck, separation, inability to speak one's truth.
Organ and related emotional imbalance	Lung: grief, loss, sadness, yearning, cloudy thinking, anguish Large Intestine: stuckness, rigidity, defensiveness, dogmatically positioned, crying
Related physical imbalance	Lung: ▪ throat/trachea soreness, vocal cord problems, cough; ▪ sweating, shortness of breath, difficulty breathing; ▪ bronchitis, asthma, emphysema; ▪ shoulder, elbow, or wrist pain, enlarged thumb joint; ▪ tightness of diaphragm; ▪ diarrhea, constipation, colitis Large Intestine: ▪ irritable bowel syndrome, intestinal noises; ▪ abdominal pain, swelling, constipation; ▪ teeth, nose, or sinus problems; ▪ neck stiffness; ▪ bursitis, shoulder or forearm pain, tennis elbow; ▪ stiffness of index finger, hand pain or weakness
Disruptive patterns create loss through	Loss of the ability to transform

Alchemist to help him give birth to his dream. He asked the Pioneer if he would be willing to go out into the world and discover a place where the dream would be free to grow.

The Pioneer was excited about this new endeavor and started his work, but soon realized he could not do his job without the Wizard's passion and excitement. The Wizard ignited the dream by enrolling thousands in the new idea, but he soon realized that he needed to call on his friend the Judge for support. The Judge then created balance in the plan by making sure that there was an equilibrium between work and play and that everyone pulled their own weight.

"We are healthiest and most vibrant when all five elements and archetypes are in balance, working together and supporting one another."

The Judge was wise and knew he needed the help of his friend the Alchemist to help transform the old way into the new way. Thus the Alchemist went to work to set in motion the process of taking the dream out into the world. After a time, the Alchemist asked his friend the Philosopher to make sure he didn't stop looking at the big picture so the dream would continue to grow and evolve. It took all five friends, supporting one another and never losing sight of their interdependence, to change the world for the better.

As in this example, we are healthiest and most vibrant when all five elements and archetypes are in balance, working together and supporting one another. If they do not work in balance and harmony—if one element or archetype is too strong or another too weak—the system not only becomes out of kilter, but instead of supporting each other, the elements weaken one other. This relates to another aspect of the five-element theory known as the "control cycle," which is based on the synergy of opposite elements.

The Control Cycle of the Five Elements:

Water controls fire because water can dampen fire or even put it out.

Fire controls metal since fire can melt metal.

Metal controls wood because metal can cut or pierce wood.

Wood controls earth because wood can block or contain the earth.

Earth controls water because it can dam water.

The Five Elements, Their Archetypes, and Associated Organs

In the traditional Chinese five-element theory, each element, or form of energy, progresses in a clockwise cycle known as a "birthing cycle"— water birthing wood, wood birthing fire, and so on. We are healthiest when all five elements are in balance, working together and supporting one another. If they do not work in balance and harmony, the elements can weaken one another. This relates to the "control cycle" of the five elements, shown by the arrows connecting the elements.

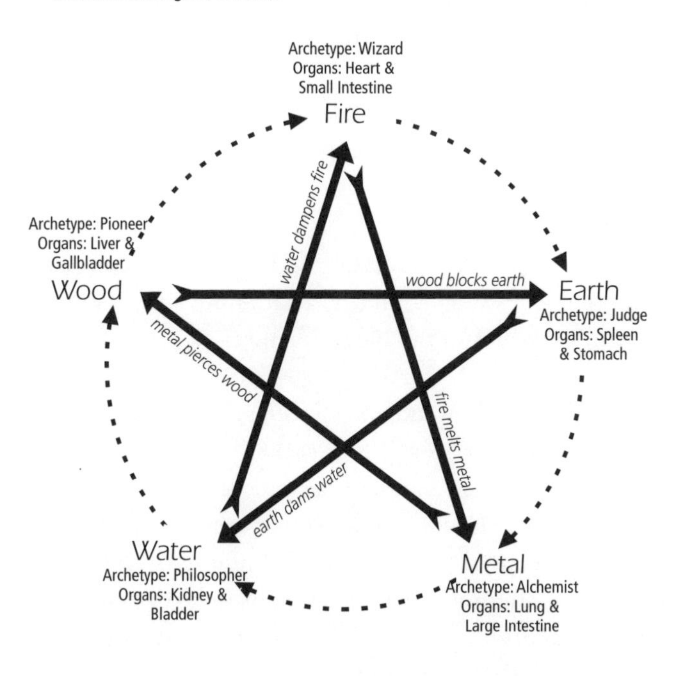

The archetypes derived from the traditional Chinese five elements provide symbolic language that helps us interpret the stories written in our genetic code.

Based on the control cycle, we can see that when the five friends—the five elements—are not working in harmony, they can create havoc in each other's lives by radiating their negative elements, which can look like this:

- The Philosopher (water) when radiating fear can penetrate the Wizard (fire) energy, dampening and putting out the fire of joy and passion that the Wizard needs to survive.
- The Wizard (fire) when radiating loss of passion and joy can penetrate the Alchemist (metal) energy, melting the metal and the ability of the Alchemist to change things, creating grief and loss.
- The Alchemist (metal) when radiating grief can penetrate the Pioneer (wood) energy, piercing the wood and the pioneer spirit of creation, which sets up anger that stops the flow of creation.
- The Pioneer (wood) when radiating anger can penetrate the Judge (earth) energy, blocking the Judge from sustaining balance and quality, which fosters feelings of not being supported.
- The Judge (earth) when radiating lack of support can penetrate the Philosopher's (water) energy, damming the water or the Philosopher's visions of the bigger picture.

As an example of how the unbalanced archetypes can play out in our daily lives, consider the story I told earlier about jeopardizing my relationship with my teenage daughter because of my inherited beliefs about what a good mother should be like. At that time, I was acting out the role of Judge, which is what I thought a good mother should be—someone who judges her children's actions as right or wrong and enforces the rules for her children's own good. It is not that the Judge (or any of the archetypes) are bad, but when they are not working in balance and harmony, we see their unbalanced expression. The unbalanced side of the Judge rigidly enforces rules. On the other hand, its supportive, healthy side creates balance (like the balanced scales of justice).

I was expressing the unhealthy side of this archetype in my relationship with my daughter. The role of the rigid Judge that I was acting out had been handed down to me through my family lineage. I learned this role from my mother, who had learned it from her mother, and so on. It was a belief system locked in my family's DNA code. I was automatically falling in line with that pattern until I finally decided to take the role of the supportive Judge, who creates balance in relationships. As I let go and stepped back, choosing to give advice and act as a guide but not to control, my daughter came forward and took responsibility for her life, and we became friends again. By changing my beliefs and behaviors I supported her deep knowingness and empowered her to realize her own greatness.

Uncovering the Emotional Issues behind Physical Problems

I once worked with a woman, Suzanne, who used the five-element archetypes to help alleviate her physical problems, which, as she discovered, were rooted in emotional patterns. Suzanne had knee pain and her kidneys were weak from the stress of her hard work. A mountain of strength for her family, she was not only responsible for earning the income to support three children, but she also maintained the house and choreographed the children's schedules as well. As a result, she didn't have much energy left over to care for herself. She often appeared hardened and even a little crusty—the exact condition of her knees. Her body was mirroring her emotions.

In Chinese medicine, weak knees are often associated with weak kidneys. I knew that if her kidneys became strong, her knees would grow stronger. We went to work looking for the emotional issue, which would be related to a story in her DNA that was weakening her kidneys. Based on the sequence of numbers she picked, that emotion turned out to be fear. Underneath her strength, Suzanne had a deep fear—a fear of failure and a fear that if she did not keep pushing, everything would be lost. Now we needed to find out what kind of inner story was setting up this pattern in her body.

The numbers Suzanne chose also led us to her operative archetypes.

Her archetypes were the Pioneer (from the wood element) and the Philosopher (from the water element). She was clearly exhibiting the unbalanced expression of the Pioneer archetype, relying too heavily on her own strength and driving herself until she was depleted. When an element is out of balance, it robs energy from another element in order to survive. In her case, her Pioneer archetype was robbing energy from her Philosopher archetype, causing the Philosopher not to be able to see the bigger picture or to understand why she was driving herself. If the Philosopher archetype had been in balance, Suzanne would have seen that she needed to slow down and she would have accepted that it was okay for her to slow down. These imbalances affected Suzanne's outlook as well as her body. The Philosopher is associated with the water element, which draws its energy from the kidneys. When the kidneys lose energy, they often weaken the knees, which is exactly what happened to her.

"Her father's and her ancestor's stories both pointed to the same theme—the loss of a pioneering spirit."

As we followed the trail of Suzanne's inner story, some important information surfaced. Suzanne revealed that when she was a child, her father had suddenly lost his long-time job as a regional manager. One day the company just closed its doors and he found himself without a job. This unexpected shift sent her father into a deep depression. He lost his Pioneer drive to forge into new fields and create new things. "Everything will be lost now because your father has lost his spirit," she remembered her mother saying. As a young child watching what happened to her father, Suzanne made a promise to herself that she would always work hard and never lose her drive, even at the cost of her own well-being.

In addition to that story, which came from her own life experience, we looked at the stories that had been handed down through her family lineage. Again, Suzanne chose a series of numbers in answer to specific questions. These numbers revealed that an ancestor who had a gypsylike nature had been put in prison and forced to work. This ancestor was given a choice of working or being put to death. The ancestor finally died in that prison from the loss of the pioneering spirit coupled with being worked to death. Her father's and her ancestor's stories both pointed to the same theme—the loss of a pioneering spirit. Both stories had created in Suzanne a fear that drove her to stay tough, push herself, and keep working no matter what.

After understanding the source of her patterns, Suzanne continued her healing process by neutralizing the energy locked in the unhealthy patterns. She eliminated the old patterns from her DNA memory and affirmed a new pattern—that she had the right to meet her own needs. As a result of going through this process and gaining a new understanding of her inner dynamics, she gave herself permission to take care of herself, she was better able to take care of her family, and both her kidneys and her knees soon became strong again.

A Loss of Passion for Life

At one of my introductory workshops, a young woman, Renee, volunteered to be part of the process I was teaching. She said she wanted to have "spiritual sight" and when I asked her what spiritual sight meant to her she said, "I want to see the world through the eyes of passion and love." Using the archetypes described in this chapter, I took her through the DNA self-healing process to see what memories in this life and what memories handed down through her ancestry were blocking her sight.

We discovered that the genetic weakness at the root of this problem had been handed down through her father's side. The numbers she picked also showed that ten generations ago someone on her father's side had lost the passion for life and had moved into a state of hopelessness. At that time, the person adopted a strategy in order to survive that required giving up the passion for life and fitting into the beliefs and standards set by society.

Renee also chose numbers that correlated to the three archetypes of the Judge, the Pioneer, and the Alchemist. The shadow side (unbalanced side) of the Pioneer believes not only in the necessity of forging new fields, but also that it takes hard work and training to achieve the desired goal. The shadow side of the Alchemist believes that the mind alone will solve the problem. The shadow side of the Judge sees and judges the world as either right or wrong. Putting all this together, the story that was unfolding revealed that someone on Renee's father's side had been rewarded for the ability to work hard and use the mind to solve problems and that this ancestor was judged and rewarded for hard work and mental ability.

Another group of numbers Renee picked revealed the next significant piece of the puzzle. It showed a pattern from her ancestry stored in her cellular memory of being a holocaust victim along with memories of poverty and punishment. Holocaust victims feel as if they are trapped or cannot change their situation, which creates the feeling of hopelessness. All of us have memories recorded in our DNA of holocausts or atrocities. A memory of poverty indicates a belief that there is "not enough." In this case, the poverty involved not having enough love or support for the passion of life. The genetic memory of a holocaust victim along with poverty set up a program wherein Renee felt as if life were a punishment and that she would be punished if she did not follow the rules and regulations of society.

After we had decoded all of this information, the young woman shared how accurate these interpretations were. She told us that her father was an engineer whose beliefs reinforced the shadow side of the Alchemist (solving problems only through the mind) and she was an athlete, reinforcing the shadow side of the Pioneer (believing that it takes hard work and training to reach our goals). All of her life, she had indeed been rewarded for her mental abilities and her ability to work hard. She was not encouraged to see the world through the Wizard's eyes of love and passion.

Through the step in the healing process called the change of heart (covered in chapter 5), she saw how this had closed down her right brain, the part of the brain where miracles and magic live. She realized that she had been taught not to trust that part of herself, to rely on her left brain for survival, which cut her off from a world filled with passion and love. This loss of passion for life caused her, in turn, to lose her spiritual sight. We completed the process by applying a technique to remove the old memories from the DNA and reinstate the memories of passion, joy, and love so that she could celebrate the beauty of life using both sides of her brain.

Several months later, I saw Renee again. She was glowing. She told me she was now living a life she had dreamed about but had never thought was possible to enjoy. She said that she no longer relied purely on her mental abilities to create her life. Once she gets the facts from her left brain, she turns them over to her right brain to see how it feels. In this way, she is able to wed her thinking world with her feeling world, restoring the sense of spiritual connection in her life.

5

The Energetic Support System

Along with the system of the five elements, another gold mine of information that helps us pinpoint the emotions and unhealthy beliefs encoded in our DNA is the ancient system of the chakras, or energy centers. The seven chakras are wheels of energy within us that support our body's energetic system. The energy flowing through our chakras is what nourishes and supports our physical body. Just as we need food, air, and water to survive, we need energy flowing through the chakras to support our life. Blockages in any of our seven chakras weaken the function of our energy centers and the flow of energy through them, in turn affecting the areas in our body that are nourished by our chakras.

What creates blockages in our chakras is the same thing that impacts our DNA—the energy patterns of our nonsupportive beliefs. Our chakras, like our DNA, hold memories, and memories that do not support us weaken our energy system or even shut it down. As long as there are no limiting patterns in a chakra, it can function unhindered and through it we can access a unique form of higher knowledge.

The chakras are usually numbered one through seven, starting with the root chakra at the base of the spine and moving to the crown chakra at the top of the head. Each one has a positive quality and expression

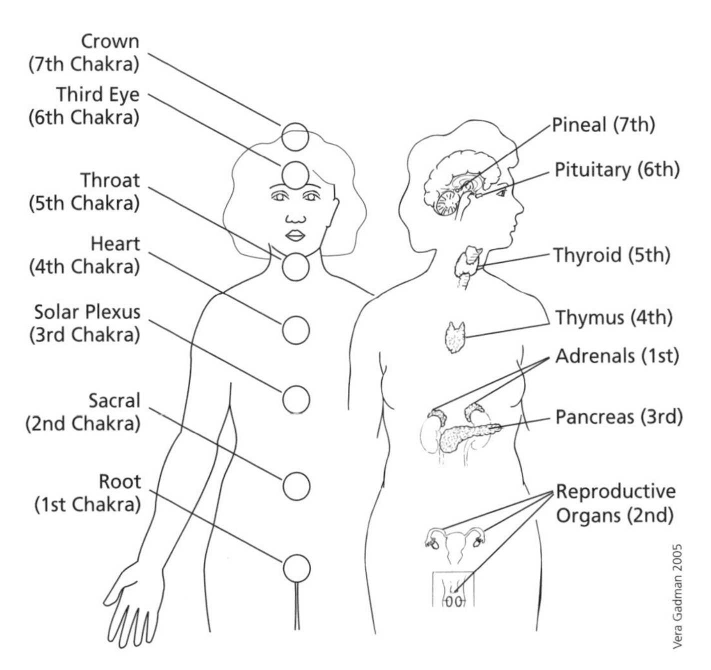

associated with it as well as a nonsupportive archetype that develops when the energy is blocked or out of balance. The charts on the following pages show these and other correspondences for each of the seven chakras.

First Chakra: Root ✪ Innocence ✪ Archetype: Magical Child

Through the first chakra, we learn how to create security and safety in our lives. When this chakra is in balance, we know how to fulfill our daily needs, which creates a feeling of being supported. Through the root chakra we can access the knowledge of innocence (or purity), where we see the world as safe, providing, and protective. When the root chakra is blocked, we feel insecure about money, work, and relationships. We feel alone, unworthy, unconnected, unwanted, and afraid of change. We subscribe to the Matrix belief that life is a struggle and that everything and everyone is a potential threat, which causes us to lose the freedom of flexibility.

The Seven Chakras in Our Energy System

1st Chakra—Root

Location	Base of spine
Gland	Adrenal
Purpose	Food/shelter, foundation
Higher knowledge we access through the chakra when blocks are cleared	Innocence or purity, where we see the world as safe, providing, and protective
Positive archetype	Magical Child (Innocence)
Nonsupportive archetype	Victim
Positive, balanced expression	A deep connection between heaven and earth. Feeling secure in the essentials needed for survival and a sense of security in the world, including physical security and financial security. Strong emotional ties with family and friends, a sense of community. A deep sense of nurturing of self, knowing that all one's needs are met, taking the time to care for one's needs. Seeing the world through the eyes of innocence, gentleness, and compassion. Creates a sharing in life's abundance.
Shadow side: Disruptive patterns create loss through	Experiencing self at the mercy of others or outside forces, creating the loss of innocence and a loss of the ability to take responsibility.
Associated element and five-element archetype	Water Philosopher

2nd Chakra—Sacral

Location	Reproductive organs
Gland	Reproductive glands

Purpose	Sexuality, creativity
Higher knowledge we access through the chakra when blocks are cleared	Creation and nurturing, the energies of the universal mother energy that create life and provide and nurture that life
Positive archetype	Goddess (Creation)
Nonsupportive archetype	Martyr
Positive, balanced expression	A sense of abundance, well-being, pleasure, loving and enjoying life. Giving self permission to create prosperity and abundance. A deep knowing of what will enrich one's life, contentment, able to see the pleasure in all things. Equal energy in all relationships (feeling that one is physically or emotionally giving as much as one is getting). Feeling that one is deserving and worthy of great things, comfortable with abundance, prosperity, and success, resourceful, rejoicing, and fulfilled. Trusting one's body, listening to one's needs.
Shadow side: Disruptive patterns create loss through	Being filled with suffering for self and others, creating the loss of the ability to sustain happiness.
Associated element and five-element archetype	Wood Pioneer

3rd Chakra—Solar Plexus

Location	Solar plexus
Gland	Pancreas
Purpose	Individualism, self-esteem, personal power
Higher knowledge we access through the chakra when blocks are cleared	Wisdom through direct connection to ancient teachings of wise ones and universal life force that gives ability to take creations out into the world
Positive archetype	Sage (Wisdom)
Nonsupportive archetype	Egoist

Positive, balanced expression	Feeling empowered and energized in self-esteem, self-confidence, and personal power, self-respecting, decisive and confident, honoring self, valuing who one is. Relationships full of quality and an equal exchange of energy, kindness, and respect. Strong sense of who one is and of truly deserving the very best. Feeling free to choose love, joy, and happiness. Creates the ability to support others unconditionally.
Shadow side: Disruptive patterns create loss through	Not acknowledging the power of anything other than self, creating the loss of wisdom and flexibility.
Associated element and five-element archetype	Earth Judge

4th Chakra—Heart

Location	Heart
Gland	Thymus
Purpose	Joy, love
Higher knowledge we access through the chakra when blocks are cleared	Love as a deep love for self and all living things
Positive archetype	Enchanted Lover (Love)
Nonsupportive archetype	Wounded Lover
Positive, balanced expression	Receptive to love, harmony, and balance within self. Loving bonds with others, open heart, kindness, forgiveness, compassion, love as the center of one's life. Love for self first and foremost, love for people, animals, plants, and all life, feeling content within self, happy to be who one is and not feeling judged. Radiance, joy, unity, kinship, experiencing the beauty and the light of others. Attracting and drawing people, cherished by children and elderly. In touch with feelings, has a profound sense of the moment. Creates kindness, forgiveness, and compassion.

Shadow side: Disruptive patterns create loss through	Not being able to protect oneself from the pain of love, creating the loss of love and the ability to sustain intimate relationships.
Associated element and five-element archetype	Fire Wizard

5th Chakra—Throat

Location	Throat
Gland	Thyroid
Purpose	Communication of highest potential
Higher knowledge we access through the chakra when blocks are cleared	Truth as the ability to speak without the judgments or teachings of the Matrix
Positive archetype	Liberator (Truth)
Nonsupportive archetype	Actor
Positive, balanced expression	Bridging mind and heart. Creativity, self-expression, willpower, truth, effective communication, enhancing one's creative spirit. Feeling of aliveness, ability to harness all energy needed for expressing feelings and wants, speaking up and being heard. Integrity, honesty, providing a safe environment to be who one is, speaking both one's mind and heart, others value what one says. Expressing creativity, truth, and integrity, knowing that one is love and that through the expression of self one's beauty shines. Creates determination and courage.
Shadow side: Disruptive patterns create loss through	Not being able to say what one wants, creating the loss of speaking one's truth.
Associated element and five-element archetype	Metal Alchemist

6th Chakra—Third Eye

Location	Third eye (middle of brow)
Gland	Pituitary
Purpose	Clear sight
Higher knowledge we access through the chakra when blocks are cleared	Clarity, the ability to see the world clearly without the distortion of the duality of the Matrix
Positive archetype	Seeker (Clarity)
Nonsupportive archetype	Analyst
Positive, balanced expression	Physical and emotional well-being. Clear thinking without illusion, great discernment, intuition, and wisdom, awakened psychic and artistic gifts, imagination as inspiration. Ability to receive guidance with awareness at all times, a bridge and connection between focused left brain and intuitive right brain. The gift of inner sight and a trust in inner knowing, a deep understanding of life and the Matrix traps. Talents as therapist, healer, artist, actor. A deep understanding of symbols, an awareness of synchronicity. The wise one. Creates wisdom and optimism.
Shadow side: Disruptive patterns create loss through	Overintellectualizing everything, creating the loss of clarity and the loss of the ability to hear one's inner self.
Associated element and five-element archetype	Earth Judge

7th Chakra—Crown

Location	Top of head
Gland	Pineal
Purpose	Faith

Higher knowledge we access through the chakra when blocks are cleared	Divine connection, a deep sense of connection to all things
Positive archetype	Angel (Divine Connection)
Nonsupportive archetype	Separatist
Positive, balanced expression	The summit of beauty, refinement, and spirituality, open to the light of spirit and eternally connected to spirit, open to receiving love and wisdom from the higher self, filled with the energy of the divine. Mastery on the spiritual plane, essence of love and awareness, linked to the oneness that resonates with the continuous flow of universal love. One with Source, one with truth, connected to all souls. Joy, peace, freedom, serenity, gratitude, knowing that all past experiences have brought one to this moment of love and acceptance. Creates spirituality as an integral part of life.
Shadow side: Disruptive patterns create loss through	Seeing all things as separate from one another, creating the loss of connection to the divine source.
Associated element and five-element archetype	Water Philosopher

Second Chakra: Sacral ✪ Creation ✪ Archetype: Goddess

Through the second chakra, we learn how to nurture ourselves. When this chakra is balanced, we are in touch with our needs and desires and we accept all our feelings, the good and the bad, rather than repressing or judging them. A healthy sacral chakra allows us to nurture ourselves while nurturing others. Through the sacral chakra, we gain the knowledge of creation and the knowledge of nurturing as we access the energies of the universal mother that create life and provide and nurture that life. When this chakra is blocked, we have a tendency to ignore or avoid our feelings. We are out of touch with our hearts' desires and buy into the Matrix belief that it is not acceptable to feel, losing our freedom to discover who we are.

Third Chakra: Solar Plexus ✪ Wisdom ✪ Archetype: Sage

Through the third chakra, we learn how to become an individual. With a healthy solar plexus chakra, we are able to stand in our power and stand up for ourselves. We value ourselves. We have the ability to take our creations out into the world. Through this chakra, we can connect to the ancient teachings of the wise ones, elders, and universal life force. When this chakra is blocked, we come from the place of the ego, which believes in control and struggle. We feel separated from our inner master, who is connected to all things. We give away our intention, our choice, and our will. Conforming to the reality of the Matrix, we lose the freedom to stop what dishonors us.

Fourth Chakra: Heart ✪ Love ✪ Archetype: Enchanted Lover

Through the fourth chakra, we learn to love ourselves. When the heart chakra is healthy, we have compassion for ourselves and we are able to forgive ourselves. We care about how we affect our family, partners, coworkers, friends, and others and we want to touch them in a nurturing way. A blocked heart chakra causes us to be hard on ourselves and to judge ourselves. When the heart chakra is out of balance, we operate by the Matrix belief that it is okay to give love but not to accept love, which causes us to lose the freedom of intimacy.

Fifth Chakra: Throat ✪ Truth ✪ Archetype: Liberator

Through the fifth chakra, we express who we are. When this chakra is in a healthy state, we are able to say exactly what we are feeling and thinking without compromise and without judgment. We understand the power of our words and we communicate with integrity. If the throat chakra is blocked, we have a hard time expressing what we want and how we feel. We speak based on the words and feelings the Matrix sees as acceptable. We give up our powers to negotiate for what we want and surrender to the power of the system.

Sixth Chakra: Third Eye ✪ Clarity ✪ Archetype: Seeker

Through the sixth chakra, we see who we are. When this chakra is healthy, we see our world as it truly is. We see how we played a part in creating circumstances where we may feel stuck or blocked. If this chakra becomes blocked, we begin to see the world through the eyes of a victim and through the eyes of duality—of right or wrong, good or bad. When it is out of balance, we experience the world through the eyes of the belief systems of the Matrix rather than through the freedom of our insight and imagination.

Seventh Chakra: Crown ✪ Divine Connection ✪ Archetype: Angel

Through the seventh chakra, we feel connected to who we are and to a higher source. When our seventh chakra is healthy, we never feel separate from our self or from our divine wisdom. When it is blocked, we experience an inability to connect to our inner knowing and the divine energy of creation. If this chakra is out of balance, we feel trapped in the Matrix and separated from the force of creation. We lose the sense of being an active partner in cocreating our world.

The chakra system is similar to the five-element system in that all the chakras are interconnected—each chakra affects the others and they support one another. For example, when you experience the world as a safe place through the first (root) chakra, it supports the second (sacral) chakra so that the world becomes a safe place for you to take care of yourself. As you take care of yourself, you are able to stand up for what you believe in through your third (solar plexus) chakra—you feel free to be an individual. As you stand up for yourself, you also value and respect yourself, creating a deep sense of self-love, supporting the fourth (heart) chakra. As you love and care for yourself, you are able to tell others, through the fifth (throat) chakra, what you need. As you speak of your needs, you are able to see, through the sixth (third eye) chakra, what you truly desire. When you truly see what you desire, you know that you are connected to a divine source (through the seventh, or crown, chakra) and can manifest your own reality.

In addition, because each chakra is connected to and affects all the other chakras, limiting beliefs that are stored in the lower chakras affect the limiting beliefs held in the upper chakras. For example, if I am holding on to a fear of survival in my first (root) chakra, it can play out in the fourth (heart) chakra as the inability to be openhearted and loving to others or to myself, in the fifth (throat) chakra as the inability to ask for what I need to survive, in the sixth (third eye) chakra as seeing the world through eyes of lack, and in the seventh (crown) chakra as the loss of the ability to create my world the way I want.

Because the Matrix has its strongest hold in the three lower chakras, the beliefs and patterns stored there must be released in order for us to access the wisdom of the higher chakras. Once the nonsupportive beliefs in the lower chakras are cleared, our energy is able to move into the heart chakra and our ties to the Matrix can be broken.

Moving into the Heart

Many of the patterns I have discovered that form the map leading to the stories in our DNA have come from watching my grandchildren grow and evolve. They showed me, for instance, that when we keep the first three foundational chakras free of nonsupportive beliefs, we move naturally into the heart, where all transformation takes place.

Unlike many of us who were raised with the common belief that you need to let children cry so they won't become "spoiled," my daughters have raised their children so that their first chakras are built on a strong foundation of safety. The children are always fed, held, and changed whenever they want to be. When our family realized that by doing this my daughters were supporting the children's first chakra, we all decided to try an experiment. We agreed to try and keep the grandchildren as free as possible of the limiting influences of the Matrix by not making them feel bad about who they were and by keeping influences that stifled their self-expression from becoming programmed into their DNA and their chakras.

One Christmas, all three of my grandchildren were visiting us and we could see through their behavior what chakras they were focusing on. My four-month-old grandson, Stryder, was developing his first chakra. The first chakra is all about survival, and he was mainly interested in being fed, held, and changed. When his base (root chakra) needs were met, he was a happy camper. It was easy for us to keep Stryder free of the Matrix influences by making sure he felt safe in the world.

Our grandson Ethan was just over a year old. Ethan was raised as Stryder had been and therefore had a strong foundation in his first chakra. He was moving into his second chakra, which is about nurturing and feeling. We made the house as safe as we could so that Ethan could explore his new world without the fear of being hurt. He put things in his mouth and we allowed him to express fully how he felt. Most of us were not raised this way. We were told that boys were not supposed to cry and we were told not to put things in our mouths. Most of us learned that it is selfish to nurture ourselves and that we are supposed to give and to take care of others first. But when we are not allowed to feel, and are not allowed to feel our own needs, the development of our second chakra becomes stunted. We become numb. In fact, a second chakra that is shut down can cause addictions. Addictions, whether to people, drugs, alcohol, food, or other things, are just a way of trying to feel. So we allowed Ethan to develop his second chakra without imposing on him the false belief that it is not okay to feel or nurture yourself.

Our third grandson, Alex, was just turning three and had moved into his third chakra, the solar plexus chakra, which is all about individualism. While it was relatively easy to keep Stryder's and Ethan's chakras open, we were not prepared for the challenges of the third chakra. Alex wanted everything his way. His favorite word was *no* and he was aggressive about getting what he wanted.

"When we are not allowed to feel, and are not allowed to feel our own needs, the development of our second chakra becomes stunted. We become numb."

Our culture does not really know how to deal with individuals. Most of us were raised with the motto "fit in and conform." So our challenge with Alex was to keep him safe and at the same time not make him feel that something was wrong with him or that he was "bad" if he did something that was not acceptable. Our big test came in a confrontation between Alex and Ethan. In his stage of life, Ethan was exploring and the one thing he was most interested in was Alex and Alex's toys. If Ethan got too close, Alex, now in the stage of individualism, would push Ethan away. While we were all sitting in a circle and visiting, Ethan once again approached Alex and, before we could intercede, Alex gave Ethan a big shove and sent him flying to the floor. In response, Alex's dad forgot our experiment for a moment, picked up Alex, and said, "I do not like you when you act like that!"

Alex held his breath. He had never heard these words before and had never been exposed to the idea that he could be disliked because of something he had done. The rest of us held our breath too, wondering if our experiment could really work. Alex's dad quickly realized what he had done, got down on his knees beside Alex, and said, "I didn't mean that I do not like you. What I mean is that I am afraid that Ethan will get hurt when you push him and I would like to find another way for you to deal with Ethan and your toys." With this, Alex stopped holding his breath and said to his dad, "I do not want you to ever yell at me again." Since Alex was able to express his desires and his individualism, we knew that his third chakra was still open.

Our patience continued to be challenged as we tried to find ways to keep Alex Matrix-free while preserving his individuality. My mother thought we should give up the experiment and go back to the old way of disciplining a child, but we decided to keep our experiment going while Alex played out his individualism to the hilt over the next six months. During this period, which we now call "Alex's Rebellion," Alex was difficult at nursery school and did not like sharing. As a result, his mother lost many of her friends among the other mothers. Then one day when she arrived at school to pick up Alex, the teacher told her that she wouldn't believe what had happened that day. She explained that when Alex arrived in class that day, the first thing he did was to see that everyone had something to play with. At snack time, he served everyone first before serving himself. No one could believe the change in him.

This was exactly what we hoped would happen if Alex's first three chakras remained free of the influences of the Matrix: *Alex had moved into his heart.* He was still a strong individual, but he was also heart-based. The greatest reward for all of us came when his parents attended their parent-teacher review. The teacher told them that she had chosen to schedule Alex's review last because she knew she would cry at the meeting. She told them that, in all her years of teaching, she had never met a child who really knew who he was and had such a deep heart for life. "When he creates something in class, he does not create it to get approval from others, but he creates things that enhance who he is," she said. "He is a strong individual who now shares his talent with others through the joy and beauty of his heart."

Most of us were not allowed to develop our individuality. We got the message that if we wanted to survive, we had better fit in and conform to the rules and regulations of society. What happened with Alex is a testimony to how we can all move into our hearts when our first three foundational chakras are free of the programming of the Matrix, when we are free of the false beliefs we inherited and hold on to. In my DNA work, a great part of my focus has been to call the real individual in each of us home so that we can access the wonders of our hearts.

Clearing the Blocks to Positive Self-Expression

Because of the way most of us were raised, the process of moving beyond limiting patterns and moving into the heart takes some catching up, and I have found that, typically, it works differently in men and women. When a woman is born and her umbilical cord is cut, her first chakra connects and grounds itself into the earth and she is in what I call the stage of innocence. Somewhere around the age of the first menstrual cycle, women energetically move into their second chakra, which is concerned with nurturing and creation. For a time, the nurturing stage feels wonderful and natural and we like to care for others. When I was in this stage, I always had enough casseroles, cakes, pies, and cookies prepared and in the freezer to feed an army so that anyone who stopped by could be nurtured. I loved taking care of our house and my children, and I loved supporting my husband.

When I entered menopause, my focus and my energy shifted to my third chakra, the solar plexus chakra, which emphasizes taking your work out into the world. As I related earlier, suddenly I was no longer interested in nurturing everyone around me. I felt I had things that I wanted to do. At first this was a strain on my marriage, but I am blessed to have a partner who was willing to work with me on this. We created a new relationship based on me doing more of what I wanted to do. Steve supported me and I supported Steve.

During this same period, I was working with a company that presented seminars on personal growth. My partner was a man. When we first started working together, I was still in the nurturing stage and wanted to support others, including him. This worked well for a time. When I shifted into my third chakra, I began creating ideas for new programs, one of which was a program to help women move through the stages of life. My partner, however, said he wasn't interested. One day we were at lunch with a group of women psychologists in Houston, Texas, and one of the women asked if we were going to go ahead with the program on women's stages of life. When my partner answered that he would be teaching it, all the women burst out laughing and said in unison that I should be the one to teach it. The next day, he fired me. He no longer knew who I was or how to handle a change in our original relationship. My independence scared him. At the time, I was devastated, but, fortunately, I remained focused on that chakra and didn't retreat back into a mode of being a nurturer only.

In this stage, when women want to assert their independence, they often become afraid that the man in their life will leave them and look for a younger, more nurturing partner (and he sometimes does). As a result, many women never move to the level of their third chakra for the fear of being thrown away, and therefore they never take their light out into the world. In order to move beyond this fear and fully develop who we are, we have to identify and then remove the programs in our chakras (which are also stored in our DNA) that are no longer serving us.

> *"When women want to assert their independence, they often become afraid that the man in their life will leave them. . . . As a result, many women never move to the level of their third chakra . . . they never take their light out into the world."*

For men, the process works differently. Men tend to move from their first chakra directly into their third chakra. Because of the general programming that takes place in our culture, men tend to skip the second chakra altogether. If and when their second chakra opens and the nurturing archetype comes alive, their focus can shift to taking care of others. When Steve started to develop his second chakra, I was not prepared for the change in our relationship, as I said earlier. Up to this point, he had been very assertive in running the business aspects of our life. Then all of a sudden he wanted to stay home and take on more of the role of nurturer. This scared me because I was afraid that if he did not want to support the family financially, we would go broke. As we adjusted to our new roles, we found that we could grow, evolve, and support each other in new ways. As in Steve's case, we found that once individuals are able to bring the second chakra to the fore and into balance, they can use the many skills they have learned throughout life in a much more effective way.

Now that Steve has brought his second chakra more into balance, he uses the nurturing aspects of this energy in all aspects of his life. In the business world, he runs our company with great success, blending his talents as a businessman with gently caring for and meeting the needs of our team and teachers. He is the best partner, father, and grandfather anyone could ask for, for he has learned how to blend his skills with his heart. He is fully present, he listens with a compassionate, open heart, and he also knows how to stand in his power and take action when needed. While I still nurture my family, I am also teaching and bringing my ideas out into the world. Had we not recognized and allowed these changes to take place in one another, our relationship would have been destroyed and we would not be enjoying the many beautiful opportunities that have opened to us since that time.

Whether you are a man or a woman, you have the opportunity to experience the positive energy of the second (sacral) chakra, which is creation, and of the third (solar plexus) chakra, which supports that creation. When all your chakras are open and clear, you have no blocks to positive self-expression through those chakras. The energy is free to move and express according to its original pattern. When this is the case, you can see and celebrate the world through the eyes of innocence, through your root chakra; you can nurture and support your creations through your second chakra; and through the third chakra, you are able to be an individual and stand in your power.

You have stepped outside the judgments of the Matrix, and with this new clarity you now have access to the light of the upper chakras. The more your first three foundational chakras are clear of limiting beliefs and limits to self-expression, the more you are able to love yourself and love others for exactly who they are through your fourth (heart) chakra. You are able to speak your truth in the fifth (throat) chakra, see the world clearly from your sixth (third eye) chakra, and feel connected to all things from your seventh (crown) chakra.

6

Neutralization and the Healing Power of the Heart

"We create our own reality." It's a saying you have probably heard many times, but what does it really mean—and how do you make it happen? Quite simply, it means that everything that is wonderful in your life you have created—*and* everything you do not want in your life you have also created. We are magnets, and we attract into our world the circumstances, people, and outcomes that match the vibration we put out into the world.

All matter is energy, and energy carries vibration. Everything on earth, animate or inanimate, carries a unique vibration. Rocks have a vibration, trees have a vibration, the earth has a vibration, and each person has his or her own vibration. In fact, just as we can be recognized by our handwritten signature, we can be recognized by our unique vibratory signature. If we are putting out a high vibration and frequency, we attract into our world things that carry that vibration, perhaps joy, health, good relationships, or abundance. If, on the other hand, we are putting out low vibrations, we will attract challenging experiences, such as anger, depression, remorse, fear, jealousy, or distress.

Life is fun when we operate at high vibrations. It is like being on a winning streak even though we may not know why. Have you ever noticed that when you fall in love, other things in your life seem to flow easily? Your car works, your days on the job go smoothly, and everything around you is wonderful—and that's really happening. I call that stage of relationship "enchantment." It is the stage we should and could live in all the time, but we don't because we eventually fall back into the ways of the Matrix. The Matrix reinforces that there are many things wrong with us, so that we naturally begin to look for what is wrong with ourselves, with our partner, and with our relationship. As a result, our vibration drops and the enchantment disappears.

Vibrations and Health

Learning how to change the vibration we are putting out is one way we can be in control of our world because by changing our own vibrations, we automatically change the vibration we attract to us. A little boy I know named Percy is a perfect example. Percy lives in England and he and his mother attended one of our courses there before the September 11, 2001, terrorist attacks in New York City. After the towers had been hit in the United States, a friend of Percy's, Oliver, showed up at Percy's door crying. Oliver was about four years old and lived next door. Percy asked Oliver why he was crying and Oliver told him, "Because my Dad said there must not be a God if something like this could happen." Since they both went to Catholic school, this was bad news.

Percy took Oliver under his wing. "It looks like you have a rather low vibration, Oliver," he announced. "Let's just fix that right now. You lie down here on this couch. I want you to feel that low vibration that there is no God." Oliver cried some more as he thought about what his father had said. "Now we're going to change it to a high vibration," said Percy. "I want you to think about ice cream, the soccer game coming up, and my birthday party this weekend and how you'll you get to sleep over." Oliver perked right up at these happy thoughts. Percy said, "Yes, that's a nice, high vibration. Now you're healed," and off they went to play.

In a very basic way, Percy understood the real dynamics of health: Our own high vibrations promote health, vitality, and well-being, and our low vibrations create the opposite. We tend to think that such a powerful transformation could not be so simple. That is because we have never been taught that disturbances in our lives (whether they are physical, mental, or emotional) are related to our energy levels. We have never been taught to understand that when we keep our body in a high energetic vibration, we won't become ill, or if we do become ill, our body will be able to repair itself. We didn't grow up learning that when we operate at a low level of vibration, we open the door to experiences that match those vibrations.

In addition to our own vibratory patterns, other people's vibrations can affect how we feel. When we are around someone with a low vibration, we tend to feel drained or disheartened. In contrast, when we are in the presence of someone with a high vibration, we tend to feel enlivened and joyful and we thrive and evolve. The goal of healing is to shift our vibrations to a higher, more positive level and to support others in doing the same. That is how we change the world—one person at a time.

"The goal of healing is to shift our vibrations to a higher, more positive level and to support others in doing the same. That is how we change the world—one person at a time."

Energy Interference Patterns

We all have an inherent blueprint of wellness and abundance. Disease is not our natural state of being. You can think of the diseased state as a condition in which our vibratory rate is out of balance with the natural, harmonious frequencies of our own body and energy field. In short, the cells in a healthy body vibrate with a high vibration and the cells in a sick body vibrate with a low vibration.

In its perfect form, DNA has a high, fast vibration. Our negative emotions and our fears form a long, slow vibration. No two vibrations can remain in the same space at the same time. Thus, when we allow a long, slow (or low) vibration into our energy field, that vibration intermixes with the high vibration of the perfect DNA and transforms it. As a result, the overall vibration of our energy field is lowered.

The pattern that is created when two or more waves of emotions ripple through each other and are out of phase is an "energy interference pattern." For example, your natural emotion of joy has its own unique wave pattern. When something happens that makes you angry, a second, low-vibrational wave pattern is created. The interaction of

The Power of High and Low Vibrational Thoughts

High vibrations promote health, vitality, and well-being, and low vibrations create the opposite. When any two vibrations meet, they create a new energy pattern. Your natural emotion of joy is a high vibration, for example. If you allow something to upset you, a second, low-vibrational pattern enters the scene. The interaction of the two vibrations produces a lowered vibrational pattern in your DNA. The goal of healing is to shift our vibrations to a higher, more positive level and to support others to do the same.

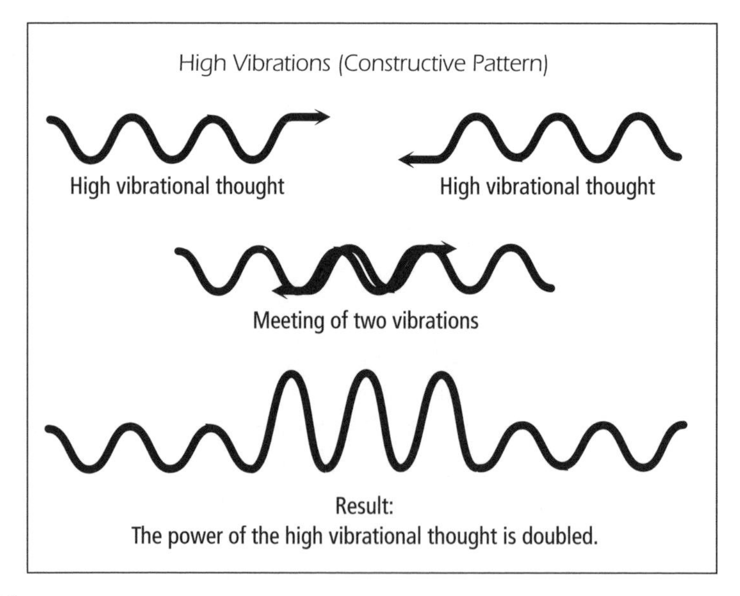

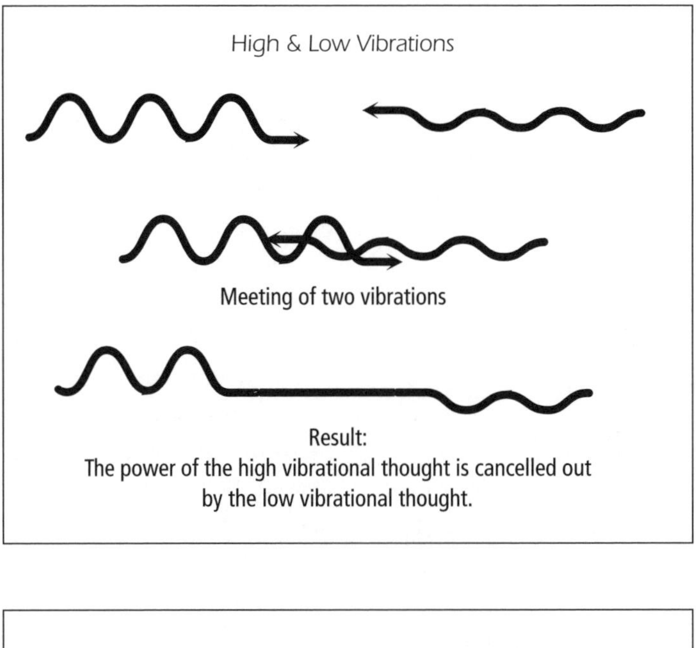

Result:
The power of the high vibrational thought is cancelled out by the low vibrational thought.

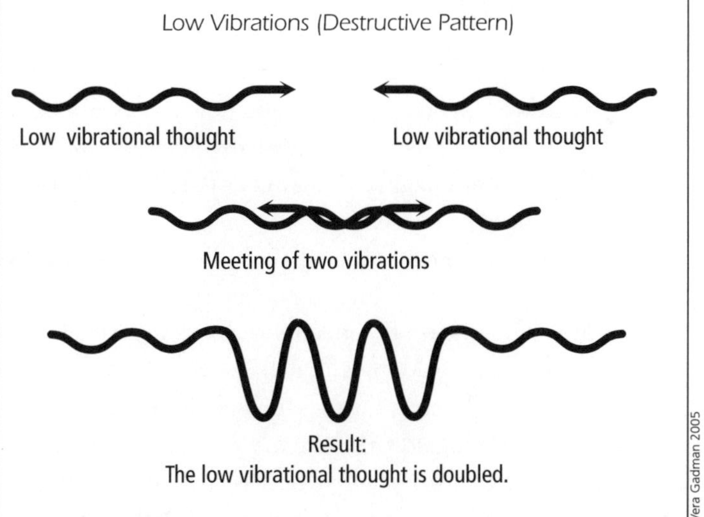

Result:
The low vibrational thought is doubled.

these two wave patterns creates a low-vibrational energy-interference pattern in your DNA, and the low vibration distorts and changes the vibration of the original DNA. If a low vibration is sustained, the flaw is distributed and it eventually compromises your immune system, causing disease or emotional imbalance. In other words, low vibrations can change a perfectly healthy genetic blueprint into a defective one. On the other hand, when you accelerate your vibrations, the low vibrations do not have a chance to distort your DNA.

We cannot, however, permanently restore our higher vibrations until we remove what is at the core of the low vibration—the programs that have been handed down through our family's lineage. In reality, weaknesses at any level are low-vibrational energy patterns held in place by limiting beliefs. Flawed belief systems upset the healthy vibrations of our cells and cause all physical and emotional imbalances. Therefore disease does not start in the physical body; it starts with the vibration of our beliefs, beliefs that have been handed down to us and beliefs we ourselves choose to sustain. Likewise, healing does not begin by looking at the disease but by working with the vibration of the disease.

"Disease does not start in the physical body; it starts with the vibration of our beliefs, beliefs that have been handed down to us and beliefs we ourselves choose to sustain."

That is why in order to heal or change your life, you must first discover the underlying, often unknown, limiting beliefs you may be holding on to. You must find the low-vibrational story that was handed down from your ancestors through your genes that has created the interference pattern in your DNA and that is causing the low-vibrational disturbance in your physical or emotional world. When the low vibration is neutralized and removed, the destructive pattern is transformed into a constructive one, allowing the body to return quickly to its inherent blueprint of wellness.

Sometimes we carry patterns for so many years that we wonder if we can really let go of them easily. Since everything we are working with is energy and vibration, it doesn't matter how long we have had the patterns or how old the story creating the vibrational pattern is. Vibrations can change instantly; and the moment the vibration changes, the DNA patterning changes as well.

I was once asked by an experienced vibrational healer if we really needed to access the stories in our DNA in order to heal or if accessing the vibration of the disturbance is enough. I explained that, in my experience, in order to heal you must remove the disruptive vibration that was created through your life experiences and the life experiences of your ancestors. Identifying the stories helps you connect to the feeling or experience of what took place. The story lets you get in touch with the vibration of the event that originally set up the energy interference pattern in your DNA. Although all healing is ultimately done through vibration, when you are engaging in self-healing, it can be difficult to get in tune with the correct vibrations. Discovering your DNA story through the DNA self-healing process has made this task easy. By identifying your ancestral story and understanding how that story has affected your life, you connect with the feeling of that story (its vibration)—and the feeling of the interference pattern at the root of your DNA story *is* the exact vibration that needs to be released in the healing.

Returning Energy to Its Natural State

Identifying the belief patterns that underlie disease (using numbers and archetypes as I explained in chapter 4) are just the first steps in the DNA self-healing process. Before we can heal and before we can manifest our hearts' desires, we must also *eliminate the core pattern from our DNA programming* that is at the root of our disturbance and *change the emotional charge* that is feeding the pattern. I call this process *neutralization.* There is no lasting healing without neutralization. If we continue to carry the vibration of limiting belief patterns, we will continue to repeat those same self-defeating patterns. In order to let go of those limiting patterns, we first need to neutralize the energy that is behind them.

All energy is neutral in its natural form. By itself, energy does not possess a concept of good or bad, right or wrong. Energy is not discriminating;

it simply moves where we direct it. The energy of a nuclear reaction can be used to create electricity or to make an atomic bomb. What determines the shape our energy takes? Intention. Energy takes on the form of our intentions, words, thoughts, emotions, and beliefs, creating the environment and the reality within us.

When we see the world through eyes of anger, for instance, we begin to experience the world as an angry place and then angry people are attracted by that vibration into our lives. The negative patterns we hold affect not only our energy field but also the energy fields around us. Unless and until we change that emotion of anger, our DNA code will remain the same and we will repeat the same pattern. The process of neutralization breaks the vicious cycle. Neutralization changes the emotional charge of the energy, returning it to a neutral state. Once energy is in a neutral state, we can redirect it to manifest a new state of being. The key to restoring energy to its natural, neutral state is stimulating a change of heart. When you know how to work with heart energy, neutralization and transformation can take place easily.

"Once energy is in a neutral state, we can redirect it to manifest a new state of being."

The Healing Power of the Heart

Within the center of your being is your sacred heart space. This is the place where all transformation takes place, where all old memories and patterns of disturbance can be neutralized. When you are centered in your heart, you do not look outside yourself for answers and you do not follow the will of someone else when deciding what is best for you. Tapping into your heart energy creates a path of aliveness where you can respond with compassion to the needs of the moment rather than relying on beliefs formed in the past to make judgments. The neutrality and perfect equilibrium of the heart allow you to move beyond judgments to see the larger possibilities before you.

Returning to the example of what happens when we first fall in love can help us understand how this works. When we become enchanted with a person's every thought and action, we are seeing from the perspective of our heart and we are experiencing that person's true essence. Without our filters of judgment and duality, we are seeing the person as he or she really is. Although this is a feeling we desperately try to hold on to—for it reminds us of our own inherent majesty—most of the time we let the judgment of what we feel is right or wrong creep in and we lose sight of the other person's highest potential. When that happens, we no longer recognize the magnificence of the person we are in love with. We can no longer hear the secret language of the heart, the language of perfection; we only hear the data and language of our beliefs.

"Tapping into your heart energy creates a path of aliveness where you can respond with compassion to the needs of the moment rather than relying on beliefs formed in the past to make judgments."

As long as we continue to see life through the filter of our past experiences and have expectations built on those experiences, it becomes difficult to see our way out of the labyrinth. Once we have a change of heart, however, we gain what I call "innocence perception," which allows us to penetrate the veil of duality and illusion and to see beyond our limitations and our judgments to a new reality.

Removing your judgments is absolutely key to changing your DNA. You can try to set your emotions aside, you can stuff them, but that does not change them, because their vibrational pattern is still there. To change your DNA, *you must move from an emotionally charged state to a state where the judgment that created the emotion in the first place is no longer there and therefore the emotion can no longer be triggered.* Accessing your heart space gives you the power to do this by helping you move into a state of compassion and love with complete detachment from the situation.

The healing power of the heart is not theoretical. Researchers at the Institute of HeartMath have shown that the heart's energy field has measurable effects on other processes in our body, on our physiology, on our DNA (as I discussed in chapter 1), and even on those around us. They have demonstrated that "brain rhythms naturally synchronize to the heart's rhythmic activity, and also that during sustained feelings of love or appreciation, the blood pressure and respiratory rhythms, among other oscillatory systems, entrain to the heart's rhythm."[1]

Furthermore, their research has shown that positive emotions such as appreciation or love are associated with heart rhythm patterns that are smooth, ordered, coherent patterns and that negative emotions such as frustration and anger correspond to disorderly, erratic, incoherent patterns. New research, they say, suggests that the "heart's field is directly involved in intuitive perception."[2] Reviewing earlier research from the Institute of HeartMath, Richard Gerber wrote in his landmark work, *Vibrational Healing,* "It appears that love may actually be a real energetic force, not merely a catalyst for action, transformation, and healing."[3]

In my work, I have found that love is the preeminent healing force. When you center in your heart and consciously access the energy and rhythm of the heart for healing, you are setting up a field of equilibrium that brings all things back into balance.

Creating a Change of Heart

The neutralization technique of the DNA self-healing process evokes the healing potential of the heart. It allows you to create a change in heart so you can let go of the pattern that is no longer serving you. Through a change of heart, you see the old pattern from a new perspective. You begin to realize that no matter how bad, painful, or disruptive the pattern or experience, it still served you in some way. All who were involved in your pattern were teachers. They were in your life to awaken something within you and to help you evolve. Perhaps they modeled what you did not want in your life or what you did not want to become. Perhaps the pain they created in your life helped awaken you to compassion, forgiveness, or understanding for others and for yourself. In other words, they helped you become who you are today. For neutralization to take place, you must be able to recognize the lesson the old pattern afforded you and you must be able to change your perspective to see the disturbance, or the people associated with the disturbance, as a gift to your evolution.

Here are some examples of the gifts we can gain from our difficulties. The pain we may have experienced as children from the loss of a parent can open our hearts to others who are experiencing a loss. It may help us model compassion. Those who were beaten or abused as children often awaken to the gift of empathy and they gain the ability to assess if a situation is safe or not. Those who had difficulty standing up for themselves often learn how to be flexible and adaptable. Those who used control to keep from getting hurt often develop leadership and will. Someone who preferred to hide behind the scenes because they were afraid of being scrutinized may gain the ability to stand back and see the bigger picture. Those who avoided making waves to sidestep conflict may now have the gift of creating balance and bringing resolution to a situation.

By acknowledging that every pattern has served a purpose (although it may now have outworn its stay), you can see the old pain from a new perspective, making it easier to let go of the patterns and move on. Since the heart is the place where all energy vibrates in a neutral state of love and deep calm, it is from the heart that you are able to see the old patterns from this new perspective and bring them back to their original, natural state of love.

Entering your sacred heart space also helps you create another essential component of healing: forgiveness. Not being able to forgive means that you have adopted a set position on something you do not entirely understand. When you withhold forgiveness, you create a burden on the heart, which carries the heaviness of grief and loss from the abandonment and betrayal you are holding on to. Many of us have misconceptions about forgiveness. To forgive someone does not mean that you were "wrong" and they were "right" in a certain situation or that you are giving in to someone or something and therefore going against what you believe in. To forgive someone does not mean that you are weak in heart. Forgiveness actually supports and strengthens your heart. It means you are willing to bring resolution to the situation and it releases you from the burdens you have been carrying.

When you are in your sacred heart space of neutral, unconditional love, you create a new environment where it is no longer difficult to forgive. As you will see in chapter 8, the DNA self-healing process first helps you discover the gift that the people related to the disturbance instilled in you. Next it helps you cultivate the specific change of heart that will help you see the old pattern from a new perspective. Then it leads you through a process of forgiving anyone, including yourself, who you feel has caused you pain or held you back. Within your heart, you tell each person how you were feeling about them or about the situation, that you now understand that the source of the disturbance between the two of you came from old patterns handed down through the ancestral lineage. You tell them that you see how the disturbance between the two of you has given you the strength and talents you would never have had without the pain of the past. Finally, you tell them that you are letting go of the old pattern, the pain, the hurt, the grievances, and the judgments so that you can both move on. Telling them this opens the opportunity for you to receive a message from their higher self to you.

"To forgive someone does not mean that you are weak in heart. Forgiveness actually supports and strengthens your heart."

Changing Our Relationship with Ourselves

The neutralization technique is not a one-time thing but a process we can use over and over to reach a deeper level of understanding and healing. I've been teaching this technique for years and I still find it indispensable to learning from and moving through new issues that arise in my life. Recently, for example, my family had a reunion to celebrate my mother's birthday. My mother and my two sisters live in California and my brother and I live in Idaho. It had been several years since we had gathered as a family and we were all looking forward to the occasion. When we first were together, we were filled with excitement and anticipation of the days ahead, but the more time we spent together, the more we started falling back into the old family patterns and roles that had still not disappeared even after years. We started to play out our old childhood strategies with one another.

As the oldest child, I found myself taking on the role of organizing everything and everyone. One of my sisters adopted her old strategy of taking a less dominant role so she wouldn't be rocking the boat. My other sister reverted to her pattern of fighting to be heard and recognized. And once again, my brother became invisible in the conflict, at times injecting humor to break the tension in the air. It hurt me deeply to see the old pain in their faces—the pain of trying to figure out who they were and what was their position in the family. On top of it, my mother announced that she was leaving California and moving to Idaho, turning everything upside down. My younger sisters immediately felt as if they were being abandoned. The more abandoned they felt, the more dramatic the strategies became. Although I had nothing to do with my mother's decision, my sisters saw me as the instigator. I felt as if I were being crucified—and resurrected.

The feeling of crucifixion did not stem from their blaming me for taking our mother away from them. It came from my not liking the role I was playing and feeling that I was powerless to change the situation. The feeling of being resurrected came from my awareness that this state, this old pattern and the beliefs of who I was in my family, had to—and could—move to a higher level. I realized that the real nail holding me to the cross was how I felt about myself. I saw that the real problem for my brother and sisters and me was that the fear that had surfaced because of my mother's upcoming move had blinded all of us to the possibilities. Instead, we were seeing the situation through our own limitations.

I knew that no matter how much we would like others to change, we only have the power to change ourselves. This put the entire situation back on me. If the present situation were going to change, I needed to focus my attention on what I truly desired rather than focusing on what I did not want. I realized that if I neutralized the situation within myself, it would shift the attention away from the present world of disruption into a world of expansion where I could have the relationships I truly desired with my brother and sisters.

First, I needed to forgive myself for falling back into the roles and strategies of my childhood. I had to let go of the beliefs of who I was in the family. As I did this, I started to see the situation from a higher perspective and from compassion. I could feel the confusion in the heart of my sister who had taken the less dominant role. She was a strong mother in her own family, did not understand why she did not have a leadership position in this family, and felt as if she were shrinking among her siblings. Her higher self was reminding me to stay strong and true to my heart's desires. I could feel the pain of my baby sister and her desperate need just to be recognized and heard in a family that was strong willed. Her higher self was reminding me that we must first hear and understand ourselves before we can be heard by others. I could feel my brother's pain as he used humor to detach from what was happening. His higher self was reminding me of the art of observation and detachment.

I consciously drew the situation and the feelings surrounding the situation into the center of my heart and asked that the old patterns that were causing me pain be neutralized. My siblings might choose to continue in their present roles, but I was no longer going to be attached to their limitations. I could view what was happening from a state of compassion. I made a conscious shift in perspective. Before, when the strategies of my siblings had triggered my feelings, I became resistant. My resistance, in turn, created energy for my siblings to react to. Now, because I had neutralized the old pattern within me, my sisters' resistance stopped. They were still unhappy with the news of the move, but they were no longer reacting from a state of anger. At that point, we were able to discuss how we felt versus simply reacting.

When the emotions of a DNA pattern are triggered, we react to the memory of the feelings and emotions of that pattern. Once the feelings are neutralized, we and those involved in the situation can move on rather than being stuck in an endless cycle of resistance. We tend to resist because it creates traction and we mistakenly confuse traction with power. We think that by holding on or by resisting, we can exert influence. You can understand how this might work in the context of a committee meeting where the committee leader is presenting ideas everyone likes and is getting all the attention. If someone wants to be noticed or to stop the flow of the meeting, he starts resisting and soon everyone is paying attention to him. While he has certainly created traction by resistance, he has not created true power. Traction and resistance create fear, worry, procrastination, judgment, and denial. If we want to move forward to the next level, we must neutralize our resistance.

"If our relationships are to heal and evolve, we must change our relationship with ourselves."

The family reunion taught me once again, this time at a deeper level, that awareness of a problem is the first step to healing, but it is not enough to change a situation. If our relationships are to heal and evolve, we must change our relationship with ourselves. We need to rise above the limitations that the problem has created. We need to move beyond seeing the problem through our present reactive feelings, which makes us feel limited and trapped, to seeing with compassion. We must recognize that our hearts can lead us to a place of unconditional love where we can let go of our judgments and see with new eyes what is really taking place.

Love is the highest frequency or vibration there is, and our DNA in its perfect form resonates and vibrates with the feeling of love. Unfortunately, many of us have forgotten what it feels like to receive unconditional love. We have been taught how to give love, but we have not been taught how to receive love and so we are hard on ourselves. Because we are afraid of being abandoned or rejected, we beat ourselves up to remind us to be good so that we will not be abandoned. We often take on the belief that we deserve to be punished and we even see illness as a good form of punishment. We not only judge others who are ill and wonder what they have done wrong, but we also judge ourselves. When we can move past this limiting belief that illness or misfortune is "bad," we can see the circumstances in a new light. Illness can become a teacher, a way of helping us move out of our limiting beliefs. Illness, therefore, is never a mistake but a way of helping us evolve.

I believe that we all came to earth with a great plan and that we made that plan in heaven before our time on earth. We made a plan that would help us grow and evolve in what I call earth school. Sometimes the plan of our life feels so terrible that we can't believe a plan like that could have been made in heaven! Yet illness has a role in the context of this plan. Illness is telling us that somewhere we have taken the wrong turn and it is time to get back on track. It is the body's way of telling us to wake up and pay attention to what is not working in our lives and to what is taking us away from the vibration of our perfect self.

"Healing is not just about 'fixing' physical, emotional, and mental imbalances. Healing is understanding your life's journey and gaining the ability to move forward with that journey."

When an illness occurs, it may be a signal to sit back and evaluate why this situation happened and what aggravated or triggered the illness. What is the original story residing in your DNA that is not compatible with the vibration of your perfect blueprint of wellness and abundance? How did you come to accept and allow that belief to perpetuate itself? How has it shaped your actions and reactions and created a circumstance you are not happy with? Healing is not just about "fixing" physical, emotional, and mental imbalances. Healing is understanding your life's journey and gaining the ability to move forward with that journey. It is about keeping your heart open to whatever lessons come your way. Once we understand that our life has purpose, we can accept and embrace all lessons we encounter as part of a positive growth process, without judging ourselves or others who bring us our lessons.

Taking an Active Part in Creating Your Own Reality

When we use vibrational healing techniques like the neutralization technique, we become a director of energy. We are connecting with the infinite supply of energy from our Source. We are directing our DNA communication system to raise the low vibration of a chaotic pattern within the body to a higher one of health. Our DNA naturally wants to vibrate at a high frequency and we are in control of whether or not it does.

It comes as a surprise to many people that we are in control of our health and what happens to us. A good part of the time, knowingly or unknowingly, we give our power away to others and play the role of the victim. Victims give the power to create their reality to someone or something outside of them. When we begin to view situations we are uncomfortable with as learning experiences, and acknowledge that we have the ability to change them, we are making the shift from victimhood to consciously taking an active part in creating our own reality and controlling our own destiny.

A major part of my life's mission is to help people move past the idea that they have to look outside themselves for answers and to teach people to heal themselves. If I were just a healer, your healing would be dependent on me. Instead, I see myself as a teacher. While there are times when we need the support and help of a practitioner, I always encourage people to learn how to heal themselves because by learning to self-heal, you have so much more power. In fact, you already know how to heal yourself. We all know how to heal ourselves. We've just forgotten how to do it.

"We all know how to heal ourselves. We've just forgotten how to do it."

The scientific revolution and new discoveries about the energetics of DNA are transforming our culture's old belief system of looking outside ourselves for healing to an inner knowing that we have the power to heal. In just the past 144 years, medicine has gone through a dramatic revolution. Less than 150 years ago, during the time of the American Civil War, we moved away from the practices of leeching, bleeding, and purging. From the 1860s to the 1950s, a period that Dr. Larry Dossey calls the era of materialistic or physicalistic medicine, Western medicine developed the main treatment strategies of drugs, surgery, and radiation that are common today. These are based on the laws of energy and matter articulated by Sir Isaac Newton in the seventeenth century.

In the early part of the twentieth century, medicine excelled in diagnosis but had no cures. It wasn't until World War II that effective drugs and surgical interventions came into practice. Up to the 1940s, we had no practical antibiotics. In the late 1960s, scientific exploration of the mind-body connection began to enter the mainstream. Until then, the idea that emotions played a part in the healing process was just a concept. "Chakras" and "energy medicine" were foreign ideas until just a few years ago, when people like Caroline Myss, author of *Anatomy of the Spirit,* made *chakras* a household word by sharing the concept with millions of people viewing *Oprah.* Chakras are now referred to in movies and literature and have become a part of our popular culture.

"Our world is in a desperate place right now. . . . By healing yourself, you become a major catalyst for changing the world."

Today physicists are moving beyond the boundaries of the physical plane. They have mapped every gene in the human body and are now looking at intuition and consciousness as parts of the healing process. They are coming to the conclusion that every cell has an innate intelligence that controls our health, and that we can direct the energy that tells our cells what to do.

With this revolution has also come the idea explored in this book—that our DNA is not fixed, that it can be altered. If we believe that DNA is a fixed code, as scientists used to, then we have no other choice but to look to some power outside ourselves for healing. When we are open to the possibility that our DNA can be changed and that we have the power to affect how our own body operates, we can make the shift from looking outside ourselves for healing to looking inside.

The majority in our culture still support the old belief. Many won't even consider a new way of approaching healing until they experience a major crisis, a challenge, or a betrayal involving the old belief. Often it starts with a midlife crisis, when the system we have become accustomed to no longer seems to work. It feels like a betrayal because everything that we believed in and poured our heart into is no longer working. The crisis can come in the form of a divorce, losing a job, the rebellion of a family member, or, as is often the case, an unexpected illness, when our body itself seems to be betraying us. At that point, there are no answers where we usually look for answers. This is the start of the journey inward.

Our world is in a desperate place right now and I know you have been asking yourself what you can do to make it better. What you can do to change the world is to heal yourself. When you are healed, your vibration changes and it goes out into the world. One person vibrating at the vibration of love can change the vibration of hundreds of thousands of other people. By healing yourself, you become a major catalyst for changing the world.

7

Manifesting Your Heart's Desires with Imagination and Intention

Your future is not set in stone. You possess the ability to change the upcoming story of your life—to determine your own destiny. Once you recognize that your belief systems and those handed down to you have been determining your reality and then identify and neutralize the specific belief patterns that have been limiting you, you are ready to create your new manifestation.

Chapters 1 through 3 explained how the experiences of our ancestors were solidified into beliefs that then became encoded in their genes and were passed on to us. These chapters explored how our own experiences and emotions can trigger the codes sleeping within our genes and showed how we adopt societal rules and patterns ("the Matrix") that do not always support the reality of who we are. Chapters 4 and 5 explained how we can quickly identify the patterns that are a part of our inner story by using numbers and archetypes, and chapter 6 described how we can neutralize the energy that feeds the limiting patterns by moving into the heart's field of unconditional love and forgiveness. Neutralization, which takes energy that has been locked in or blocked and returns it to its neutral state, is like washing the slate clean—it clears the way for us to reshape the energy and bring into reality our heart's desires, a process called manifestation.

Why is manifestation important? Many of us have inspiring visions and feel that we are here to do something grand. Others of us are just awakening to this possibility. No matter where you are on your path, you have the ability to make a difference in the lives of others. You can shift the consciousness of others as you yourself make a shift and as you claim your right to create your own world—to manifest.

Manifestation is the finely tuned art of aligning your mind, your heart, and your thoughts with the powerful energy of creation. Manifestation can come in many forms. You can manifest something small that you need or something big that you have dreamed. You can manifest a healthy body, a parking space, an appointment, or the money you need to pay the bills or to go on the vacation of your dreams. Manifestation can bring the right job or the right person into your life. There is no limit to what you can manifest.

You are, in fact, already creating your own world. Your thoughts and beliefs set up a template for what you are creating, and your feelings energize those thoughts and beliefs, magnetizing their essence to you physically. As you understand how this template works, you become more skilled in the art of manifestation and you start to open up to a larger world that many call the world of magic and miracles. It may feel like magic when your manifestations come to life, but what is really taking place is that you have learned how to use your thoughts and feelings to direct your energy in constructive ways.

Flexing the Muscles of Your Imagination

Each one of us is meant to be a powerful force for manifestation. Most of us, however, use only a small fraction of our brilliance and power. The two keys that can unlock that power are imagination and intention. Imagination and intention work hand in hand to guide the powerful energies of love for healing and manifestation. When intention and imagination are fused, anything is possible.

Imagination is developed in the right brain, which I call the mind's eye of creation. As children, "making believe" was part of life. We made believe we were Superman, a cowboy, a princess, a doctor or a nurse, a bird or a plane. Through dreaming and pretending, we strengthen the muscles of imagination and creativity that we need to draw on throughout our lives. Although imagination is one of the most powerful tools we have for bringing our manifestations to life, it is also one of the most misunderstood.

Imagination is of one of those words that has conflicting implications. The dictionary says that imagination is the picturing power of the mind and the creative principle of the mind. At the same time, it tells us that imagination is a fabrication and a fallacy of vision. We might say that one person "is the victim of her imagination," implying that her thoughts are untrue, and that someone else "is a person of great imagination," implying she is a great visionary. In addition, many of us have been punished for using our imagination. We were taught that having an imagination meant we were lazy and that daydreaming or making things up was a waste of time. We started to see imagination as our enemy when it is really our friend, for you can only create what you can imagine.

"When intention and imagination are fused, anything is possible. . . . We can create miracles, but we must first imagine them."

I once worked with a teenager who was trying to heal her pain of feeling abandoned as a child. I knew that before something can become a reality we must be able to imagine it. I asked her to remember a time when she felt truly loved. She looked at me and said she couldn't remember a time like that. So I asked her to imagine what it would feel like to be truly loved, and she said she didn't know how it would feel and couldn't imagine it. At that moment, I realized how devastating the loss of imagination can be. Without the power of imagination, we cannot create our dreams or remember our original wholeness.

When we were children, most of us were never told by anyone that we were the light of the world or that we were one of the most magnificent beings that had ever been born on the planet. Instead, in the hopes that we would become a better person, we were told to fit in and act our age, to grow up, for heaven's sake. As a result, the only way we can remember our real magnificence is to imagine it. With this teenager, we had to start at the beginning. I read her stories that talked of beauty and love. We rented movies that expressed the feeling of being loved for who you are. Once she was able to imagine what it would feel like to be truly loved, she was able to heal. We *can* create miracles, but we must first imagine them.

The Power of Visualization

Visualization is a key tool for harnessing the power of our imagination to create a new reality. Dr. Charles Garfield talks about the powerful effects of imagery on our physical performance in his book *Peak Performance.* He was introduced to the idea when he met with a group of Soviet sports trainers and asked to see a demonstration of their techniques of mental training for athletes. The trainers decided to demonstrate their techniques by working with Garfield himself, who used to lift weights but hadn't done serious training in years. They asked him how much he thought he could lift and then asked him to try and lift 20 pounds more, for a total of 300 pounds. Garfield barely managed to lift the 300 pounds. He was pleased but exhausted. The trainers asked him how long it would take him to be able to lift the maximum he used to lift. Since weight lifters increase their capacity in small increments, Garfield estimated that it would take about nine to 12 months of workouts to get back to 365 pounds. Then the trainers announced that by using their techniques, he could do it within an hour. Although he thought that was impossible, Garfield decided to try the experiment anyway.

First, the trainers asked him to lie down and then they used guided imagery to help him deeply relax. Next they led him through a visualization of each detail and step of weight lifting. They asked him to visualize his hands and the weights. They asked him to imagine what his breathing would be like, what his muscles would feel like as he lifted the

365 pounds and what the weights would sound like as he moved them up and then down. Garfield got up and walked over to the bench, becoming nervous. The trainers again reviewed with him the relaxation and visualization technique. Garfield says that while he was engaged in the visualization the second time, the imagery gave him confidence and that it began to guide his movements—and he went on to perform his incredible feat.

When I was teaching one of my seminars at the University of Glasgow in Scotland, a professor there was conducting research on the mind's brain-wave states. She invited me to observe one of her experiments. She brought a person into the room and had him play the piano, and as he played she monitored his brain waves. She found that he moved into a different brain-wave state, the alpha state, while playing the piano. She then asked the same person to move away from the piano, close his eyes, and imagine he was playing the piano. Just by imagining, his brain waves changed to the alpha state. The same holds true for healing: When we imagine a person or ourselves as whole, perfect, and already healed, miraculous things can occur.

"When we imagine a person or ourselves as whole, perfect, and already healed, miraculous things can occur."

To enhance your visualization of what you want to bring into your life, spend some time flexing the muscles of your imagination. You can ask yourself the following questions and even write down your answers to help anchor your vision.

First, imagine yourself already having your manifestation in your life right now. Feel the joy, the love, and excitement of having what you truly desire.

What do you see?

How has it enhanced your life and what wonderful changes has it brought into your world?

Imagine how you will look to others when they see you with your manifestation.

What is your environment like (your home, workplace, or other location involved in your manifestation)?

How has your manifestation affected your relationships?

Using all five senses, imagine exactly how your manifestation feels. What does it look like, sound like, smell like, taste like, and feel like?

Intention Directs Imagination

The next step in manifestation is to merge imagination with the power of intention, connecting the right brain (the mind's eye of creation) to the left brain, which I call the mind of certainty. The left brain is where we take responsibility for our creations and determine their blueprint. When you fuse imagination and intention, you have directed, or controlled, imagination.

Imagination + Intention = Manifestation

A research scientist from Japan, Dr. Masaru Emoto, has demonstrated the tremendous power of intent by conducting thousands of experiments with water, which he has published in *The Hidden Messages in Water* and other books on the subject. Dr. Emoto discovered that words, ideas, and music have a profound effect on the molecular structure of water. Dr. Emoto took water samples from a dam in Japan and found that its crystals were very deformed. He then asked a chief priest to pray over the polluted dam water and took pictures of the water's crystals before and after the prayer. The pictures of the water sample before the prayer showed crystals that were horribly misshapen. After the priest's one-hour prayer practice, the water formed beautiful crystalline structures that, in Dr. Emoto's words, emitted "brilliant energy."

Dr. Emoto wrote various words in different languages on pieces of paper and wrapped the paper around jars of water so that the words faced the water. He found that, afterward, the crystalline form of the water changed dramatically depending on the intent of the words. For example, words and phrases like "you fool," "demon," and "you make me sick" produced crystals that were distorted, dispersed, and ugly. When words like "love/appreciation," "thank you," or "wisdom" were taped on the jars of water, the water's crystals formed beautiful, geometric forms. In other words, the *high-vibrational words* changed the water to a pristine, crystalline form, and the *low-vibrational words* changed the water to a chaotic, random form.

In addition, Dr. Emoto found that various kinds of music changed the water dramatically. When he played heavy metal music into a container of water, the water crystals he photographed were broken into pieces. When he played selections by Bach, Beethoven, and Mozart into separate containers of water, the water crystals, although each shaped differently, all produced exquisitely beautiful forms. Again, certain kinds of vibrations lowered or raised the water's vibration. We can understand the real impact of Dr. Emoto's work when we remember that the human body is composed of 70 percent water. Just as the water in the jars changed, the water inside us—and therefore our body itself—responds to the vibration of various words and to the healing powers of intention and love.

"Intention is the spark, ignition, and the focused beam of energy that calls forth the people, places, and synchronicities we need in order to bring our manifestations to life."

Intention acts as the driving force that guides our imagination. Intention is the spark, ignition, and focused beam of energy that calls forth the people, places, and synchronicities we need in order to bring our manifestations to life. We cannot expect imagination alone, without the power of intention, to make something happen, just as we cannot expect intention by itself to make something happen. A great example of someone who fused the two was Walt Disney, who linked his child's gift of imagination with intention to manifest the hugely successful "World of Disney."

Imagination and intention work interactively with your DNA. When you imagine something, your DNA code is temporarily altered. This alone will not produce a permanent alteration or a permanent signal within your DNA until you use the power of intention to support your imagination. You could be watching a movie that makes you sad and imagine that the same sadness will come into your life. While you are watching the movie and feeling sad, your DNA takes on the pattern of the sadness, but the new pattern only remains in place while you are feeling the sadness from the movie. If, as a result of your continued attention on this movie, you keep feeling sad or you alter your life, this will affect your DNA on a more permanent basis.

Therefore, if you are just imagining that you want a new relationship or better health or a different job, but you do not have the driving force of intention behind it, your imagination alone will not be strong enough to alter your DNA code and you will not see a change in your life. Imagination and intention are able to alter the DNA because together they create a strong constructive energy pattern in the DNA. This, in effect, reestablishes the original blueprint of perfection in the genetic code, restoring it to its inherent state of wellness and abundance.

Details, Details, Details

How do you create a strong intention? By clearly focusing on your desires, you create a thought and emotion that emit a signal—the vibration of your intention—that draws to you the same vibration you are sending out. Intention requires attention to detail. Intention is the blueprint for manifesting change, and the details are what make up that blueprint. Assembling the details is like gathering the ingredients for a recipe. A good recipe lists all the ingredients and what proportions of those ingredients are required. Likewise, the recipe for your manifestation should include the features, traits, characteristics, qualities, and aspects you want in your manifestation.

In essence, intention is creating a complete and detailed picture before you light a fire under your imagination. Here are some examples. If you are manifesting a home, for instance, you first want to get a clear idea of its look and feel. Start with the feeling you want to create and answer the following questions.

How will you use your home? Will your home be a place for community, love, or family? Will it be a place for entertainment or a place to retreat in seclusion?

What kind of atmosphere will it have?
How big do you want it to be?
What will it look like and who will be living there with you?
How will you decorate the house?
Where will the sun rise and where will it set?
Will there be plenty of trees or will there be a view?
How many and what kind of rooms will your home have?
How will you entertain yourself in this home?
Will you own or rent the house, live in it full-time or part-time?
What is the house made of and what does it smell like?
What do you see yourself doing in your home?

If you want to manifest something that is not a material possession, you could ask yourself questions such as:

What exactly would I like to have that I do not have now (in the areas of career, relationship, well-being, or prosperity, for example)?
What would this give me that I do not have now?
How would this make me feel?

I have a dear friend who was living in New York and who wanted to move to the Northwest. She visualized herself moving to and living in a beautiful house on a lake in this area of the country. Within three months, she was living in a gorgeous home on a lake in Idaho. However, she had left out one detail—that she *own* the home. As a result, she was in her dream home, but as a house sitter. The universe will provide you with what you need, but *you* need to provide the details.

Revisit Your Values

In addition to defining the details of what you want to manifest, the next way to spark your power pack of intention is with motivation. Motivation comes from your values. You cannot sustain the vibration needed to bring the manifestation to life if your manifestation is not in alignment with your values. Values are the principles you live by, your standards, ethics, and moral code. To discover your values, ask yourself:

What is it that I value at this point in my life? Is it freedom, aliveness, integrity, simplicity, compassion, abundance, security, romance, or individuality? The list of what you value will be unique to you and will change as you grow. Once you are clear about your values, you can determine whether what you desire to manifest is in keeping with them. If what you value is not in alignment with what you want to manifest, you will not be sending out the correct signal to attract what you desire. Instead, you are setting up a conflict and a disconnection that will prevent your manifestation from coming into your life.

For example, say you want to manifest a perfect relationship and have decided that you want this relationship to bring more excitement into your life. Say you decide to imagine a partner who is the life of the party, but what you truly value is someone who will be attentive to you and who is dependable. Since what you are imagining and what you value are not necessarily the same thing, you have set up an energy field of confusion, and that is exactly what will show up in your life. To attract your manifestation, you must send out a clear signal.

Sometimes we have to trick the brain in order to discover what we truly value. I have found that the following process can help bypass your brain and give you answers from your heart—the place where your true values come from.

Write down your favorite movie, your favorite book, your favorite song, your favorite holiday, and your favorite person.

Now ask yourself the following questions and write down the answers.

What do you love about the movie?
How did it make you feel?
What do you like most about the book?
How does your favorite song make you feel?
What do you love most about your favorite holiday?
What do you admire about your favorite person?

You have just compiled a list of your true values. You might have answered that what you loved about the movie was its theme of *freedom,* what you liked most about the book was its concept of *individualism,* what you loved about the song was the feeling of *romance,* what you loved about the holiday was the feeling of *community,* and what you admired about your favorite person was his or her *loyalty.* Thus you have discovered that what you value most are freedom, individualism, romance, community, and loyalty. These are your true values and anything you want to manifest as well as its details must be aligned with those values.

Laser Beam Focus

The next way to spark your power pack of intention is focus—laser beam focus. Intention is related to the word *certainty*—that is, knowing without a doubt that what you want to create *will* come into being (and is, in fact, already here). The realization of your manifestation depends on your ability to imagine it with laser beam focus as already fulfilled. Visualize what you want, see it in place in your life, and fully expect it to unfold. Whenever we sidestep, vacillate, backtrack, detour, and refuse to make up our minds or to take a stand for what we truly want, we dilute or lose the power of intention.

When you want to bring something to life quickly, intensify your focus. When you focus intently on your desire and leave no room for anything else to crowd out your intention, the energy you are sending out to the universe to call forth your manifestation is tripled. Act as if your manifestation were already part of your life. Speak as if it were already here. See it happening and taking place in your world. *Know* that the new manifestation is on its way as you continue to focus on and use your imagination and intention to fuel it.

"Act as if your manifestation were already part of your life. Speak as if it were already here. See it happening and taking place in your world."

If you are manifesting a house, for example, let every house you see remind you of something you want in your own house. Don't just think about your manifestation once or even once in a while. Focus on it as much as possible. Cut out pictures of your dream house and put them up on the wall or somewhere you will look at them every day. Make a list of what you need to build or buy a new house, including the builder you want to build your home or the real estate agent you know will create the best deal. Find the area or the lot you want to build on. Spend time there and spend time seeing yourself there. Watch a sunset to see which way you want the house to sit. Start choosing the flooring and the tiles. *See your dreams as if they were already happening and act as if they have already come true.*

Whether you are intending to manifest a new job, a healthy relationship, or something material, don't waver in your focus until you get results. Think about and feel your manifestation as real—knowing it is already here. Each morning when you awake, state your manifestation and feel your manifestation as real—knowing it is already here. Each evening before you go to sleep, state your manifestation aloud and feel your manifestation as real—knowing it is already here.

Be Flexible

Although we can detail and visualize exactly what we want to manifest, we cannot know how the manifestation will appear in our life and therefore we have to stay open to how it will unfold. We interrupt the process by being specific in our instructions to the Universe about exactly *how* a manifestation should come into our lives. My daughter Allison watched this principle in action in her life when her husband was attending graduate school at the University of Michigan. She had just had her first baby and wanted to come home and visit us, but at the time she did not have the money for the trip. She was also lonely and wanted a friend since she had just moved to the area.

When I first taught her the manifestation technique, she said, "I'm going to manifest a friend," and went off to practice the eight steps of manifestation I now teach in my seminar "Ancient Secrets of Manifestation," including activating her imagination and intention. Thirty minutes after she had used the technique, a young woman rang her doorbell and said she had met Allison's husband at school that day, that she and her husband had also just moved to the area, and that they too had a new baby. She wanted to know if she and Allison could be friends.

Stunned at what had just happened, she called me. "I think this works," she said. "I am coming home this weekend. I'm going to manifest the $500 I need for a ticket." She then applied the technique to manifest the money for her ticket, not knowing where it would come from. That night when her husband arrived home, he said, "They made a mistake on my scholarship and today they gave me a check for $500." Allison suddenly had her airfare.

A single mother who came to one of our classes taught the manifestation technique to her 16-year-old daughter, who was graduating from high school and wanted a Volkswagen Beetle more than anything in the world. She had a picture of one on her bedroom wall, she had gone to the dealership and sat in one, and she knew that a Bug was exactly what she wanted. The mother was doing her best to support the two of them, but buying the car for a graduation present was out of the question. So her daughter took the situation into her own hands and used the manifestation technique to get her car. Shortly afterward, she walked to the local 7-11 and bought a bottle of cranberry juice. When she looked inside the lid, she found she was the instant winner of a brand-new Volkswagen Bug.

Removing the Blocks to Manifestation

Just as it is essential that intention and imagination work together for manifestation, it is important to emphasize again that for healing and wellness to manifest permanently, we must not skip the first four steps of the healing process. For our manifestation to be successful, we must first identify and neutralize the limiting belief systems (the energy blocks) that keep generating the same limiting patterns. The stories written in our genes affect our ability to bring forth our manifestations and can detour our dreams.

Like all of us, I have programs and stories in my genes that enhance manifestation and that block manifestation. For instance, one of the rules of manifestation is to know that it can happen instantly. That was a world I was used to. My father would be poor one day and rich the next. My father's father had seven brothers and sisters. All of them had quite a bit of money, but only my grandfather had married. As each of these relations would pass on, my father would inherit a large sum of money. As a child, I did not understand how we got the money we had. All I knew was that it was possible to get something instantly.

When we were living in a small town in the central valley of California, we couldn't afford a car so we rode our bikes everywhere. One day my father came home with an airplane. We were probably the only family in America who did not have a car but had a plane. On another occasion, my father woke all of us and said, "Call a friend and invite them to come with us because today I am flying all of us and your friends to Hawaii for the week." And he did.

While manifestation came naturally to me, in order to take my skills to a higher level I had to remove other programs that compromised my ability to manifest. Along with the empowering stories, I also had stories that were not serving me—memories stored in my genes that made me believe that we have to work hard for money, that it is difficult to hold on to what we have, and that the more money we make, the less time off we will have.

Imagination, as I said earlier, comes from the right brain and intention from the left brain. What stops our manifestations from taking place is a third area of the brain that sets up a barrier between imagination and intention. I call this part of the brain the Matrix mind. The Matrix mind tells the left brain that it is impossible for imagination to change things, that change is based on fact, and that fact is based on our five senses. As I described in chapter 3, Matrix programming wants everything to fit within its standards. It wants us to think that nothing works outside the Matrix. It can fill our minds with doubt, confuse our intention, and extinguish our hopes and dreams, for you cannot be positive and negative at the same time; one will rule the other. Matrix self-talk sabotages both our imagination and intention, and tying into the group thinking of the Matrix (the peer pressure and patterns of our culture) can slow down the materialization of our desires. To attempt to manifest your heart's desires before you remove the blocks of the

Matrix is to struggle against the nature of things. There must first be an inner change before you will see an outer change in your manifestation work.

The DNA ladder looks like a double-stranded helix consisting of two chemical strands. The rungs in the middle are made up of combinations of four amino acids that compose the genes that hold your programming. Looking at this energetically, you could say that the right strand holds the programming for your imagination and the left strand holds the programming for your intention. Imagine that you want to manifest a trip to Hawaii. You have been visualizing it, you imagine yourself lying in a hammock on a beautiful Hawaiian beach and drinking a piña colada. For your trip to manifest, the first strand of DNA (the programming for intention) must ignite the spark of imagination held in the second strand.

If you haven't eliminated the influence of the Matrix, the Matrix programming will butt in with negative self-talk that confuses your vision. "*Hello,* is anybody there?" it will bark. "You're not being realistic! Did you forget that you have to work because you haven't earned enough vacation time? Did you forget that you have zero dollars in your bank account? Do you really think you can get away with being so irresponsible? There is absolutely no way you can go to Hawaii." After hearing that, it's easy to see why we would doubt the power of imagination to bring into our life what we need. The Matrix, with its negative programming, snuffs out the spark of our imagination—and without the spark, intention dies. Forget the trip to Hawaii! To successfully merge your imagination with your intention, you must remove the blocks that the Matrix mind has set up.

I overcame a big block to one of my manifestations when I recognized how the Matrix programming was working in my life. It happened as a result of a series of interchanges I was having with my daughter. As I said, Allison had really learned the art of manifestation. She had recently manifested a remodel of her home. She loves all the details involved in remodeling, everything from designing the plans to going to the hardware store to pick out fixtures. Each morning she was calling me and filling me in on all the details. With each call, I noticed that I was becoming shorter and shorter with her, as if I had no patience to

Manifesting Your Heart's Desires with Imagination and Intention

Intention + Imagination = Manifestation

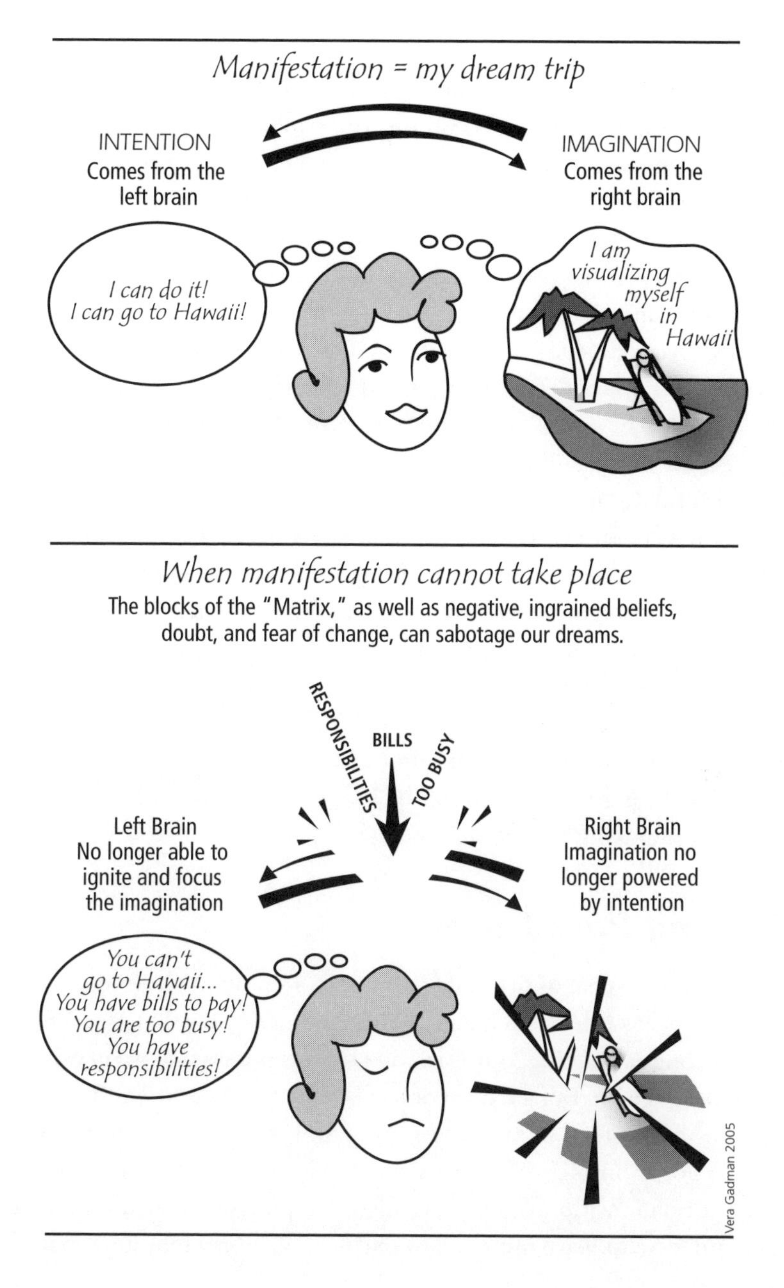

hear about the details of her new project. I did not understand why I was agitated and tried to ignore it—until I realized there was a message in it for me.

My husband and I had been in the process of manifesting a home on the lake, but my rational mind (in fact, the Matrix mind) was telling me that if someone knocked on our door today and gave us a million dollars, I still wouldn't be able to find the time to take care of the details of building a new house. I loved teaching and that required me to travel frequently, so I didn't have much spare time for anything else. I was thinking about this and I suddenly understood that my Matrix mind had been running the show. I realized that I was irritated with Allison because every time she shared the details involved in her remodeling, my Matrix mind was reinforcing the idea that my dream of a new house was impossible. It was telling me that I did not have the time or the interest for the details involved in building a new home, and therefore I couldn't have one.

I knew that if I continued along this line of thinking, I would never get a new home. I immediately went to work to remove the blocks within me through the neutralization technique. Then I called Allison and told her what had happened and why I had been so short with her. By waking up to what was happening, I also discovered the way out of my dilemma. I decided to hire Allison to help build our new house. With great excitement, I continue to do what I love—to travel and to teach—knowing that very soon our home on the lake will be a reality, with Allison taking care of the details.

Moving Past Fears

Another aspect of the Matrix that will often block the energy of our manifestations is fear of change. Sometimes we worry that as a result of manifesting something, our lives will change in ways that we may not be comfortable with. In fact, most of us fear change even when the present conditions in our lives are not working and the change would be good for us. For example, some of us worry that if we actually manifest the perfect relationship we want, we might find ourselves with a new partner and have to leave our current partner. We worry that if we manifest the perfect job, we might end up working for a new company or living somewhere else. I call these fears the blocks of the Matrix. They are the Matrix's attempt to keep things as they are.

To defeat the Matrix mind, we have to learn to identify the fears that are just part of its programmed self-talk so we can move beyond them to new possibilities. The Matrix hates change and uses fear to keep us from changing. When fears come into your consciousness and try to take you away from your focus, you can do two things to get back on track. *First, by recognizing the fear as something that supports the Matrix and not you, you can tell these fears that they have no place in your world and must leave your consciousness immediately.* Standing up to the Matrix is where the strength of our intention comes into play. Intention is the power to hold the thought and the feeling associated with your manifestation until it comes to life. In difficult situations at work, at home, or with friends and associates, you may hear the voice of the Matrix, but if you're listening closely, you will also hear the voice of intention that shouts, "Come on! We can do this!" That is the voice that will lead your team to victory. If you allow it to, the force of your intention can override the rational voice of the Matrix, and when it does, you open yourself up to the magic of manifestation.

Second, you can use a powerful technique that neutralizes fears by centering in the heart and sending the fear to God. *Envision a triangle between you, God, and the fear. Bring your attention to your heart, center your attention there, and then draw an energetic line from you to God. Then draw an energetic line from God to the fear. Next draw an energetic line from the fear (now neutralized by God's love) back to your heart.* Meditating on this energy flow allows God's energy to dissolve the fear.

For example, if you have a fear of failing a test, this sets up an energetic connection between the fear and you. The fear feeds your beliefs that you will fail the test. As long as the vibration of the fear of failure is active, you are connected to it and it will direct your energy toward failure. Energetically, a direct line from your heart is connected to the fear. You can neutralize the energy by changing this straight line into a triangle where the energy moves from your heart to God and from God to the fear. As you center in your heart and connect your heart energy with the energy of God, which is love, God takes over and sends the healing light of love to the fear of failing a test. God's healing light of love neutralizes the fear. The energy is now in a neutral state as it reconnects back to your heart and you are no longer connected to the fear of failing a test.

When we get past the fears of the Matrix and our own core patterns of fear, we return to the energy of innocence and wonder that is part of our natural world. It is this energy that opens the doors to creation. As we grow older, we tend to lose the sense of innocence and wonder. We stop seeing the world in a grain of sand, as William Blake put it, and we put our attention on how things are done and how to do them in the proper manner. As you rekindle your childlike imagination, you will find that wonder lies at the heart of manifestation.

"Liberating ourselves from the dos and don'ts of Matrix thinking . . . unleashes our creativity, our happiness, and our power. It shows us possibilities where we thought there were no possibilities."

A woman who attended one of my classes wrote me that through working with the manifestation process we teach, including the key elements of imagination and intention, she had come to realize how much the programmed patterns from her childhood had stifled her creativity and happiness. "As a child, I was often punished for making too much noise, grounded for being late coming home from school, or ridiculed for using my imagination," she wrote. "This translated in me as a fear of having fun (and maybe laughing too loud), a shutting down of my sense of wonder and beauty (don't dare stop and smell the roses—I might be late!), and a fear of being laughed at for creating anything."

She went on to say how grateful she was for the many miracles that had occurred in her life in the year since she had learned to clear energy interference patterns and recapture her innate creativity. "It has been a long time since I could honestly say, 'I'm happy,' and truly feel it in my heart. It's really amazing to be able to feel this, especially when

I had forgotten what it was like!" Liberating ourselves from the dos and don'ts of Matrix thinking with our own powers of intention and imagination unleashes our creativity, our happiness, and our power. It shows us possibilities where we thought there were no possibilities.

Part III

How to Reset Your Genetic Code

8

The Five Steps to DNA Self-Healing

The world of healing is a world of infinite possibilities. When we get stuck thinking that one way is better than another or that there is only one correct way to do something, we limit our growth. When we allow ourselves to get trapped into seeing things from only one point of view, our life becomes stagnant. Everything is always evolving, and if we can learn to move with the flow of evolution, our life will flow freely along the path that is meant for us. As our heart opens to new possibilities, we begin to notice the small things that can lead to enormous discoveries. When our heart stays closed to possibilities—when we are so sure that we already know all the answers—we may miss the great opportunity to be guided by our higher knowing, which speaks to us in simple ways and is always guiding us.

I named my company PossibilitiesDNA because the word *possibilities* reminds me that there are many ways to do things and that there are always new opportunities to evolve into. Knowing that there are always more possibilities gives me the chance to continually discover the new. It fills my life with wonder, innocence, and excitement. It has given me the freedom to thank the old ways for the lessons and opportunities they have brought me and to embrace the new with a sense of adventure and a promise of expansion.

Marlo Morgan, author of *Mutant Message Down Under,* once told an interviewer a story about a group of aborigines floating on a raft in the ocean. She asked them why they didn't use paddles to help guide and move the boat in the direction they were headed. They answered that if they used paddles, they might end up somewhere other than where they were supposed to be going. They knew the raft would take them exactly where they needed to be.

To me, that is a great example of trusting the universal flow of life. My old way would have been to paddle like mad to get where I thought I should be. Acknowledging possibilities gives me the opportunity to let the raft flow where it is supposed to go. It has let me open many doorways, meet many exciting and wonderful people, and experience a life I might have missed.

This chapter will give you an opportunity to experience new possibilities. In my training course "Eliminating Interference Patterns of DNA (EIP of DNA)," I have taught thousands of people around the world how to use this process to locate and then neutralize the nonsupportive core patterns that subconsciously influence us. You can use these same steps over and over again to identify the stories in your DNA that underlie any issue you are dealing with, to reset your genetic code, and to manifest the new patterns you want to see in your life. (I also offer more in-depth classes to help you further develop your abilities in healing and manifestation.)

Choosing Your Numbers

When working with the "Five Steps to Healing," be sure to allow yourself quiet time so you can give your undivided attention to the process, which works at levels as deep as you allow it to. In this process, you will be using the system of numerical mapping and archetypes that I described in chapter 4. The process itself is quick and simple. You don't have to force it. The numbers will do the work for you, leading you to the key elements of your story. You will find that simply picking your numbers and following the rest of the process will release the old patterns and create the healing or transformation you are looking for. So just relax and let the numbers and the words transform your life.

Before picking your numbers in each of the sections, remember to breathe deeply, center in your heart, and raise your consciousness to a higher perspective. This will help your brain waves move into a theta state, a state of deep relaxation, meditation, inspiration, and divine love. Once the brain has reached this state, you are out of the Matrix. You have created an energy field that bypasses the brain and takes you directly to the subconscious mind. Remember that when we are in our heart space of neutral, unconditional love, we create a new environment where it is no longer difficult to have a change of heart, to forgive, or to heal. The vibration of love that emanates from this place of perfect equilibrium is what heals and transforms our old patterns.

Once you have reached a meditative and relaxed space and you are ready to choose a number, be sure to pick the first number you hear, feel, or see. Remember, each number holds the vibration, or frequency, of the sentence or words that are assigned to it. Likewise, your DNA emits a certain frequency. According to the universal law of attraction, you will pick the number that matches the feeling or vibration that is stored in your DNA memory. If you hear, feel, or see more than one number, ask yourself which number is stronger or which number "lights up" more than the other. If both numbers still seem equally strong, you can use them both. Choosing a number is a process of faith. Give yourself permission to trust yourself and you will be amazed at the accuracy of the numbers you choose.

Five Steps to Healing
Resetting Your Genetic Code for Total Wellness and Abundance

Along with other factors, the family "stories" embedded in our DNA can determine if we are rich or poor, thin or fat, happy or depressed. These patterns don't change until our inner programming changes. The DNA self-healing process helps us discover the inner story underlying any issue, neutralize the nonsupportive patterns that subconsciously influence us, and manifest the new patterns we want to see in our life.

Step 1
Discover the theme of your story

How to find the overall theme and disruptive pattern that created pain, loss, or hardship and interfered with your having the life you want.

Step 2
Find your ancestral story

How to find the original source of the disruptive pattern as well as the strategy and program your ancestors adopted that were then passed down to you.

Step 3

Discover how these stories play out in your life now

How ancestral patternings and themes have repeated themselves in your life.

Step 4

Create a change of heart and neutralization

How to neutralize the emotions and feelings that created the patterns of disturbance.

Step 5

Create your new story

How to manifest your new story through the power of intention, imagination, and focus.

Five Steps to Healing: Resetting Your Genetic Code for Total Wellness and Abundance

Step One: Discover the Theme of Your Story

How to find the overall theme and disruptive pattern that created pain, loss, and hardship and interfered with your having the life you want.

Write down a disturbance or a block you are experiencing in your life. It can be in the area of health, money, relationships, career, or something else, or it can be a pattern of thinking or behavior that keeps arising.

____

____

Center in your heart in preparation for choosing a number from each of the categories listed.

To move into the heart's vibration of divine love, focus your attention on your physical heart. Think of your heart as a giant rose. Breathe deeply into your heart space and slowly release your breath. Repeat this three times. As you breathe, you are aligning your physical body with the vibration of love and you are connecting with your sacred heart space.

After breathing deeply into your heart space, take your thoughts high to the heavens. Feel yourself releasing any stress. Imagine that you are floating on a cloud. Let a feeling of lightness and deep peace come over your body. When you can feel a sense of peace or calmness, you have moved into a state of divine love, which gives you direct access to your subconscious mind. (If you don't feel that you can reach a feeling of peace right now, know that your intention alone will let the process unfold. Form the intention that you will choose the correct numbers for your story, and the correct numbers will be given to you.)

(Complete column 1 first, then go to column 2.)

1. Choose numbers to discover your story.

A. To identify the theme of your story, close your eyes, center in your heart, and choose a number from 1 to 5: ___

B. To find the energy interference pattern archetype, element, and loss, close your eyes, center in your heart and choose a number from 1 to 5: ___

C. To identify the organ and related emotion:
If you chose 1, 2, 3, or 5 in answer to B above, close your eyes, center in your heart, and choose a number from 1 to 2: ___
If you chose 4 in answer to B above, close your eyes, center in your heart, and choose a number from 1 to 5: ___

D. To find the disruptive energy interference pattern in the chakras, close your eyes, center in your heart, and choose a number from 1 to 7: ___

2. After choosing your numbers, refer to the Step One Answer Key (next page) and fill in the blanks to discover your story.

The pain, loss, separation, sacrifice, and hardships that were created through the theme of my story,

_______________________________,
(A. Theme)

have set up an energy interference pattern in my five-element archetype of

_______________________________,
(B. Archetype/Element)

creating a loss of

____
(B. Loss)

affecting my

____
(C. Organ)

and bringing up the emotion of

_______________________________.
(C. Emotion)

The pain, loss, and hardships that were created through the theme of my story have also set up a disruptive pattern in my

____
(D. Chakra)

chakra and its nonsupportive archetype of

____
(D. Nonsupportive Archetype)

creating a loss through

_______________________________.
(D. Loss)

These disruptive patterns have interfered with my having the life I want.

Step One Answer Key

A. Theme of your story:

1. Abandonment
2. Lack of trust
3. Loss of self-worth
4. Rejection
5. Betrayal

B. Disruptive energy interference pattern related to the five elements:

Element	*Loss*
1. Alchemist (Metal element)	Loss of the ability to transform old beliefs into new realities
2. Philosopher (Water element)	Loss of the ability to see the bigger picture
3. Pioneer (Wood element)	Loss of the ability to move forward
4. Wizard (Fire element)	Loss of passion
5. Judge (Earth element)	Loss of equality

C. Organ and emotion involved:

Organ	*Emotion*
If you chose #1 Alchemist for B:	
1. Lung	Grief
2. Large Intestine	Rigidity or feeling of being stuck
If you chose #2 Philosopher for B:	
1. Kidney	Fear
2. Bladder	Irritation
If you chose #3 Pioneer for B:	
1. Liver	Anger
2. Gallbladder	Resentment
If you chose #4 Wizard for B:	
1. Heart	Loss of joy
2. Thyroid-Adrenal	Confusion
3. Small intestine	Fear of rejection
4. Female organs	Suppression
5. Male organs	Suppression
If you chose #5 Judge for B:	
1. Spleen-Pancreas	Worry
2. Stomach	Lack of support

D. Archetypal disruptive energy interference pattern of the chakras:

Chakra	*Nonsupportive Archetype*	*Loss through*
1. Root chakra	Victim	experiencing the self at the mercy of others or outside forces, creating *the loss of innocence and a loss of the ability to take responsibility*
2. Sacral chakra	Martyr	being filled with suffering for self and others, creating *the loss of creation and of the ability to sustain happiness*
3. Solar plexus chakra	Egotist	not acknowledging the power of anything other than self, creating *the loss of wisdom and flexibility*
4. Heart chakra	Wounded Lover	not being able to protect oneself from the pain of love, creating *the loss of love and of the ability to sustain intimate relationships*
5. Throat chakra	Actor	not being able to say what one wants, creating *the loss of speaking one's truth*
6. Third eye chakra	Analyst	overintellectualizing everything, creating *the loss of clarity and of the ability to hear one's inner self*
7. Crown chakra	Separatist	seeing all things as separated from one another, creating *the loss of connection to the divine source*

Step Two: Discover Your Ancestral Story

How to find the original source of the disruptive pattern as well as the strategy and program your ancestors took on that were then passed down to you.

Find the weakest genetic link. Emotions from events in the lives of your ancestors have been passed down through the genetic memory you inherited. Each gene holds within its memory the first negative experience and the emotion related to it. Discovering the cause and effect of these emotions in your ancestral memory will pinpoint your weakest genetic link, which is a key to clearing the physical gene in this lifetime. The following process of finding the weakest genetic link will reveal the exact moment the defective gene became programmed into your gene, which side of the family you inherited it through, and how many generations back the defective programming began.

(Complete column 1 first, then go to column 2.)

1. Choose numbers to discover the weakest link.

A. To identify which side of the family holds the weakest link, close your eyes, center in your heart, and choose a number from 1 to 4: ___

Answer Key to A:

1. mother's mother
2. mother's father
3. father's mother
4. father's father

B. To determine how many generations back the disturbance entered the genetic line, close your eyes, center in your heart, and choose a number 1 through 8: ___

Answer Key to B:

1. 1–8 generations ago
2. 9–17 generations ago
3. 18–26 generations ago
4. 27–34 generations ago
5. 35–41 generations ago
6. 42–50 generations ago
7. 51–56 generations ago
8. 2000 b.c.–Prehistory, First Civilizations

2. After choosing your numbers, fill in the blanks from the Answer Keys to the left to discover your ancestral story.

The source of my genetic weakness came in through the lineage of my

____

(A. Side of Family)

____________________ generations ago

(B. Number of Generations Back)

and is the source of my disturbance.

Note: Each generation represents 33 years.

(Complete column 1 first, then go to column 2.)

1. Choose numbers to discover ancestral story.

A. To find the ancestral pattern that is the source of the energy interference pattern, close your eyes, center in your heart, and choose a number from 1 to 7: ___

B. To find the strategy behind your ancestral story, close your eyes, center in your heart, and choose a number from 1 to 14: ___

C. To find the program created from the strategy, close your eyes, center in your heart, and choose a number from 1 to 16: ___

D. To find the feeling that the ancestral pattern created in you, close your eyes, center in your heart, and choose a number from 1 to 11: ___

2. After choosing your numbers, refer to the Step Two Answer Key and fill in the blanks to discover your ancestral story.

The source of the energy interference pattern from my ancestral lineage that has been handed down in my DNA, generation after generation, is:

______________________________.

(A. Ancestral Pattern)

In order to survive, my ancestors took on this strategy, which was handed down to me:

______________________________.

(B. Strategy)

Because of this strategy, my ancestors took on this program, which was also handed down to me:

______________________________.

(C. Program)

The ancestral pattern set up in me the feeling of being

______________________________.

(D. Feeling)

The memories of my ancestral patterns have set up my patterns of response, which have affected my life and held me back from having the life I want.

Step Two Answer Key

A. Ancestral pattern that is the source of the energy interference pattern:

1. Failure
2. Rejection
3. Poverty
4. Betrayal
5. Punishment
6. Disappointment
7. Powerlessness

B. The strategy behind your ancestral story:

1. To survive I have to do this alone.

2. To survive I must become invisible.
3. To survive I must become aggressive.
4. To survive I must suppress my emotions using humor, diversion, or numbness as an escape.
5. To survive I must be passive.
6. To survive I must strive for recognition to ensure my worth.
7. To survive it will not be safe to be me.
8. To survive I must fit in.
9. To survive I must appear as weak.
10. To survive I must drive myself.
11. To survive I must be in control.
12. To survive I must keep myself separate from others.
13. To survive I must be loyal and dutiful to the wishes of others.
14. To survive I must keep others happy even at the expense of my own well-being.

C. The program created from the strategy:

1. Assertiveness
2. Sacrifice
3. Withdrawal
4. Resentment
5. Disillusionment
6. Perfectionism
7. Judgment
8. Inconsistency
9. Hard-heartedness
10. Obedience
11. Burn out
12. Victimhood
13. Cynicism
14. Defensiveness
15. Distrust
16. Holding back

D. The feeling that the ancestral pattern created in you:

1. Unappreciated
2. Undeserving
3. Unsuccessful
4. Unsupported
5. Unable to express myself
6. Unimportant
7. Unlovable
8. Unnoticed
9. Unqualified
10. Unwanted
11. Unworthy

Step Three: Discover How These Stories Play Out

How the ancestral patterning and themes have repeated themselves in your life.

Explore how patterns repeat. The purpose of this section is to show how the patterns that are stored in your DNA have repeated themselves in your life.

- First, center in your heart and pick any year from your birth to the present: ____.

- Take a look at the patterns you discovered from Step One: Discover the Theme of Your Story (page 147). What major event relating to the patterns you chose from Step One happened in the year you chose here?

  ____

- How did that event make you feel?

  ____

- How did that event and pattern affect your life?

  ____

- What decision did you make about yourself when this happened?

  ____

Explore the effect of ancestral patterns. The purpose of this section is to show how the present patterns and the ancestral patterns stored in your DNA affect your life.

- Think back to the year you just chose in the preceding exercise: _____.

- Take a look at the patterns you discovered from Step Two: Discover Your Ancestral Story when you chose your numbers (page 151). Which of those patterns best matches what happened in your life in the year from the preceding exercise?

  ____

- How did that event and pattern make you feel?

___

- How did that pattern affect your life?

___

- How does your ancestral pattern hold you back from having the life you want?

___

Step Four: Create a Change of Heart and Neutralization

How to neutralize the emotions and feelings that created the patterns of disturbance.

Before you can heal or change an unwanted pattern, you must change the emotional charge that is feeding the pattern. To accomplish this, you must do two things: forgive yourself or anyone else that has held you back and see the situation from a new perspective, which will create a change of heart. No matter how bad or how disruptive the old pattern, it served a purpose for your evolution. Those involved in your old pattern were teachers, contributing to your advancement. They were in your life to awaken something within you. Perhaps they modeled what you did not want in your life or what you did not want to become. Perhaps the pain they created in your life helped awaken you to compassion, forgiveness, or understanding. In other words, they helped you become who you are today. This next section will help you see how the people associated with the disturbance were a gift to your evolution.

(Complete column 1 first, then go to column 2.)

1. Choose numbers.

2. After choosing your numbers, refer to the Step Four Answer Key and fill in the blanks to discover your gift and change of heart.

A. Take a minute and think about anyone related to the disruptive patterns you identified in Steps One and Two who has affected your life in a negative manner.

To discover the gift that the person(s) related to the disturbance instilled in you, close your eyes, center in your heart, and choose a number from 1 to 14: ___

The lesson or the gift that was instilled in me from the person related to the disturbance is the gift of

______________________.
(A. Gift)

B. To discover the change of heart that will help you see the old pattern from a new perspective, close your eyes, center in your heart, and choose a number from 1 to 18: ___

My new perspective and change of heart is that

______________________.
(B. Change of Heart)

This right reinstates the feelings of freedom, love, deserving, value, purpose, and self-empowerment within me.

Step Four Answer Key

A. Gift that the energy interference has given me:

1. Compassion
2. Forgiveness
3. Understanding
4. Wisdom
5. Security
6. Safety
7. Protection
8. Self-love
9. Faith
10. Empathy
11. Gratefulness
12. Flexibility and adaptability
13. Leadership and will
14. Negotiation

B. My change of heart:

1. I have the right to feel secure and safe.
2. I have the right to be supported.

3. I have the right to my desires.
4. I have the right to feel my emotions and express them freely.
5. I have the right to care for, nurture, and pamper myself.
6. I have the right to draw boundaries and say no.
7. I have the right to stand up for myself.
8. I have the right to speak my truth.
9. I have the right to my dreams.
10. I have the right to follow my inner guidance.
11. I have the right to move forward.
12. I have the right to belong.
13. I have the right to fail and start again.
14. I have the right to follow my heart.
15. I have the right to be an individual.
16. I have the right to have relationships based on equal energy, equal support, and equal responsibility.
17. I have the right to give myself permission to stop.
18. I have the right to be seen as a precious child of the universe.

Neutralization and Change of Heart Meditation

It is now time to release the old patterns stored in your DNA and bring them to a neutral state. All energy must move into a neutral state before it can be manifested into new energy. You will be releasing the patterns you discovered in Step One (the theme of your story) and in Step Two (your ancestral story). You will also be neutralizing how these patterns affected your life. You will be focusing on your gift and your change of heart.

Remember that healing works on the principle of vibration. All disease or disturbance starts with the vibration of a belief, and healing comes about by transforming vibratory patterns. Through connecting with the vibration of love in your heart, the energy locked in the old patterns can be restored to its natural and neutral state of love. Relax and allow yourself to be guided through the following five parts of this meditation:

Relaxing. Center your attention on your heart. Think of your heart as a giant rose. Breathe deeply and slowly into your heart and then slowly let the breath go. Do this three more times. With each breath, you are connecting more deeply with the energy of your heart.

Now imagine your thoughts floating up out of the top of your head and continuing to float up to the heavens. See your thoughts resting on a cloud, and from there feel yourself connecting with God, or the universal source of knowledge. Say silently to yourself:

> I give great thanks for the neutralization of the patterns I have discovered today.

(If you have trouble taking your thoughts high, breathe deeply into your heart once more, knowing that your intention will connect you to the healing process of neutralization.)

Acknowledging the gift and change of heart. Focus your attention on anyone related to the patterns you have discovered in this process who has affected your life in a negative way. Bring to your mind's eye anyone who has held you back from becoming who you want to be or from doing what you want to do, including yourself.

Now focus on the gift that they helped instill in you (the gift from Step Four A or another gift you have received) and on your new perspective and change of heart (from Step Four B). Say:

> I am grateful for this gift of ____________________ *[name the gift from Step Four A, page 155, or another gift]*, for it has helped make me who I am today. Although the situation or circumstances of the past may have been painful and disruptive to my life, they still helped me grow and taught me key lessons. One of the lessons they taught me was ____________________ *[name the change of heart you chose in Step Four B]*, which instilled in me a new perspective. Without these circumstances, I wouldn't have developed the gifts and strengths I have today.

Forgiving. Focus again on anyone related to the patterns who held you back or created a disturbance in your world. It is now time to tell them silently, in these or your own words, that . . .

- I am now viewing the disturbance from a new perspective.
- I wish our relationship could have been different, yet I understand that the source of the disturbance between the two of us came from old patterns handed down through the ancestral lineage.
- This new information is helping me see what happened from a new

perspective. I understand now that the disturbance between the two of us has given me the strength and talents I would never have had without the pain of the past.

- I am now letting go of the old pattern, the pain, the hurt, and the grievances so that we can both move on.

Once you have silently communicated these messages, the vibration of the words will move your energy into a state of compassion and love. It is from this state of being that the old patterns will be released. Do not worry about how you are feeling at this point, for the words you have just said silently are starting to do their work at the subconscious genetic level and the healing has begun.

Once you have told the people related to this pattern whatever you needed to say, they may have a message of love to relay to you. Close your eyes for a minute and listen. You may hear them speaking to you in your heart.

Moving into the sacred heart. Your sacred heart space is where all old memories and patterns of disturbance are neutralized. Center your attention on your heart. Slowly and deeply breathe into your heart and slowly breathe out.

See your heart as a giant rosebud. With each breath you take, another petal unfolds and you move deeper into the center of the rose.

With your next breath, you have entered the center of the rose—the sacred heart. Here all energy is neutralized and returned to the vibration of love. The sacred heart knows only love; it does not judge the world, you, or others as right or wrong. Feel the magic and magnificence of this space.

Now take all the people you are forgiving, including yourself, and place them in the center of the sacred heart. Take all the patterns you have discovered today and place them in the center of the sacred heart also.

As you breathe deeply into this space three more times, the old pain and the old patterns are being neutralized.

It is done. You have neutralized the old energy.

Erasing the old story. Now that you have entered deeply into the sacred heart, the energy of its love is spreading through your body and deep into the nucleus of your cells, right into the heart of your DNA. Take a minute and feel the energy of love spreading out from your heart to the center of your cells.

With each breath you take, this radiating energy of love is erasing the old story that was stored in your DNA memory and restoring your DNA to its original state of perfection. All the memories from past generations, back to the source of your disturbance, are being released from the DNA memory.

With your next breath, it is done; your story has been neutralized. You are now ready to create your new story. Give great thanks for this neutralization.

Step Five: Create Your New Story

How to manifest your new story through the power of intention, imagination, and focus.

Now that your old DNA patterning is gone, you are ready to create your heart's desires by reshaping the energy that you have neutralized.

Set your intention. The first step in creating your new story is to move into intention. Intention is the energy of certainty—your knowing that your manifestation will take place. Intention is the blueprint for manifesting change. It is the focused beam of energy that calls forth the people, places, and synchronicities needed to bring your manifestations to life. When you state with certainty what you desire to manifest, you move into intention.

First, ask yourself: What is it that I want in my life now?

Write your answer in the blank below and be as specific as possible. Then say your intended manifestation aloud, with certainty.

My manifestation and intention is: ______________________________

____.

Feel your intention. Feeling the new story you want to create is essential to manifestation. To help you step into the feeling of your intention, complete the steps below.

Write down the element (metal, water, wood, fire, or earth) you picked in Step One B (page 147).

(Write answer in opposite column)

The element I am working with is:

__________________________________.

(Answer from Step One B)

To identify the feeling of your intention, close your eyes, center in your heart, and choose a number from 1 to 6: ___

Go to the following Answer Key and find the number under the element you are working with. Fill in the answer to the right.

As I work to manifest my intention, the feeling behind my intention is:

__________________________________.

(Answer from Answer Key)

Answer Key to Feeling of Intention

Metal—the Alchemist

1. I am competent.
2. I am perfect as I am.
3. I stand in my own power.
4. I am recognized.
5. I can create change.
6. I have the right to do what I love.

Water—the Philosopher

1. I can see the bigger picture.
2. I can remain true to my basic nature.
3. I am creative and original.
4. I no longer look outside of myself for the answers.
5. I know when to take action and when to step back.
6. I can embrace the future.

Wood—the Pioneer

1. I am free to move ahead and create excitement in my life.
2. I have great creative power.
3. I have great visionary potential.
4. I am inspirational.
5. I help others to move forward.
6. I am confident.

Fire—the Wizard

1. I am passionate.
2. I have the ability to transform things.
3. My imagination is my gift.
4. I can create miracles.
5. I am optimistic.
6. I know what I want and I am like a magnet that draws that energy to me.

Earth—the Judge

1. I create balance in all that I do.
2. I have great skills as a mediator.
3. I create a loving environment for others and myself.
4. I have great compassion.
5. I am supported.
6. I stand up for myself.

Detail and imagine your manifestation. Take a few minutes to answer the following questions and to imagine your manifestation:

What will my manifestation look like?

How will it enhance my life?

Now, imagine what your world will feel like and look like when your manifestation has taken place.

Feel intensely. An important step in manifesting is to lock into the feeling behind your manifestation. It is the feeling of what you want that will bring the manifestation to you. The following questions will help you get in touch with and generate that feeling.

If you are manifesting financial abundance, ask yourself:

- What is the money for?
- What will I do with the money?
- How will I feel when this is done?

If you are manifesting relationships, ask yourself:

- What do I want in the relationship?
- How will I feel if my needs are met in the relationship?

If you are manifesting health, ask yourself:

- What do I want to change about my health?
- How will I physically feel when I'm healed?
- How will I emotionally feel when I'm healed?

If you are manifesting a situation or a material possession, ask yourself:

- Why do I want this?
- How will I feel when I have this?

Manifestation Meditation

It is now time to create your new story. You will create your new story through the ancient process of manifestation. Relax and allow yourself to be guided through the six parts of this meditation.

Stating the manifestation. Center in your heart space and breathe deeply into your heart four times. Take your thoughts high to the heavens and connect with God, or the universal source of knowledge.

State aloud your manifestation (from the beginning of Step Five, page 159). Say:

My manifestation is ______________________________
____.

Creating the geometric field for transformation. Geometric shapes alter and help focus the subconscious field of DNA. They create a vortex of directed energy. The shape of a pyramid amplifies the energy so that it can transcend to its next level of evolution, grounding the energy into that dimension.

Visualize a golden pyramid. Now feel this golden pyramid coming down over your entire body until you see yourself standing in the center of the pyramid. You have now entered the temple of transcendence.

Releasing the interference pattern of DNA. Now that you have entered your temple of transcendence, look to the bottom of the pyramid on your left side. Imagine a ball of energy sitting in this corner. This energy contains the belief systems, the patterns, and the disturbances of your past. This energy holds a negative vibration.

Now look to the bottom of the pyramid on your right side. Imagine a ball of energy sitting in this corner. This energy contains the energy of your future, the energy of your new manifestation. This energy holds a positive vibration.

Now feel the negative vibration coming into the center of the pyramid, into the center of your being. Now feel the positive energy coming into the center of the pyramid, into the center of your being. The two vibrations cannot merge, so you may feel a little chaos at first as the two vibrations create an interference pattern at the cellular level.

Feeling the manifestation. To transcend the chaos, look to the top of the pyramid, where the feeling of what you want to manifest resides. *Feel* what you want in your life, now that the old patterns have been released. Go beyond words and *feel* what your life will be like with your new manifestation. Feel the joy, feel the freedom, and feel the peace that your manifestation will give you.

Pull this feeling down into the center of the circle, into the center of your being. As you continue to pull this feeling down into the center, the old interference is being transformed. Feel a deep sense of peace within your being as this happens. Know that your intention alone will create this change.

Give great thanks for this transformation of energy. It is done and your manifestation is created.

Sending the new energy out into the world. Once your energy has been transformed, you send it out into the world to attract to you every person, place, or thing you need for your manifestation. See this energy radiate out through the pyramid like a lighthouse. The photon light of your DNA has changed the program within you and its signal. This new light energy is a magnet, drawing to you exactly what you need. Feel everything you need as part of your manifestation coming back to you.

Give great thanks for your manifestation. Say silently to yourself, "It is done, it is done, it is done."

Maintaining laser-beam focus. Don't waver in your focus until you get results. Continue to feel your manifestation. See your manifestation as part of your world now. Speak as if your manifestation is here now, already created, already done.

As you are falling asleep at night, state your manifestation aloud and see and feel it as real, *knowing* it is already here. When you first wake in the morning, state your manifestation aloud and see and feel your manifestation as real, *knowing* it is already here.

Endnotes

1: Your DNA Storybook

1. Caroline Myss, *Anatomy of the Spirit: The Seven Stages of Power and Healing* (New York: Harmony Books, 1996), p. 40.

2. Ibid., pp. 34, 40.

3. Candace B. Pert, *Molecules of Emotion: The Science behind Mind-Body Medicine* (New York: Touchstone, 1997), p. 187. Reprinted with permission of Scribner, an imprint of Simon and Schuster Adult Publishing Group. Copyright © 1997 by Candace B. Pert.

4. Ibid., p. 189.

5. Rob Stein, "Study confirms that stress helps speed aging," MSNBC, Nov. 30, 2004, www.msnbc.msn.com/id/6613721.

2: Your Emotional Heritage

1. Gregg Braden, *Walking between the Worlds: The Science of Compassion* (Bellevue, Wash.: Radio Bookstore Press, 1997), p. 77. Reprinted with permission.

6: Neutralization and the Healing Power of the Heart

1. Rollin McCraty, et al., "The Resonant Heart," *Shift: At the Frontiers of Consciousness* (December 2004–February 2005), p. 16. See Dr. Rollin McCraty, Director of Research, Institute of HeartMath, www.heartmath.org, and *Shift: At the Frontiers of Consciousness,* Institute of Noetic Sciences, www.noetic.org.

2. Ibid., p. 17.

3. Richard Gerber, *Vibrational Medicine: The #1 Handbook of Subtle-Energy Therapies,* 3d ed. (Rochester, Vt.: Bear & Company, 2001), p. 527.

About the Author

Margaret Ruby is a leading educator and pioneer in the fields of personal growth and self-healing. She is a gifted energetic practitioner and the founder of PossibilitiesDNA, a school that teaches cutting-edge vibrational self-healing techniques and also certifies energetic practitioners.

For more than 20 years, Margaret Ruby has researched and studied how thoughts and emotions, both past and present, can alter our DNA and affect all areas of our lives. She has created successful transformational techniques anyone can use to discover and remove the nonsupportive belief systems that are causing disease and disturbances. PossibilitiesDNA has taught these techniques to thousands of people all over the world with profound and life-changing results.

Margaret regularly conducts her seminars and training courses across the United States, Canada, and the British Isles. The Georgia Nurses Association, an approver accredited by the American Nurses Credentialing Center's Commission on Accreditation, has approved Margaret Ruby's course on the energy interference patterning of DNA as a continuing nursing education activity. In addition to healing techniques, PossibilitiesDNA offers courses on relationships and the ancient secrets of manifesting abundance.

To learn more about Margaret Ruby's work, including free lectures in your area, self-healing classes, certification training courses, private consultations, CDs, DVDs, and other products, call 888-711-9962 (outside the U.S., call 208-265-1507), visit www.possibilitiesdna.com, or write to PossibilitiesDNA, 495 Sherwoods Road, Sagle, ID 83860.